Hydrogen Production Technologies

Hydrogen Production Technologies

Editors

**Suttichai Assabumrungrat
Suwimol Wongsakulphasatch
Pattaraporn Lohsoontorn Kim
Alírio E. Rodrigues**

MDPI • Basel • Beijing • Wuhan • Barcelona • Belgrade • Manchester • Tokyo • Cluj • Tianjin

Editors
Suttichai Assabumrungrat
Chulalongkorn University
Thailand

Suwimol Wongsakulphasatch
King Mongkut's University of
Technology North Bangkok,
Thailand

Pattaraporn Lohsoontorn Kim
Chulalongkorn University
Thailand

Alírio E. Rodrigues
University of Porto
Portugal

Editorial Office
MDPI
St. Alban-Anlage 66
4052 Basel, Switzerland

This is a reprint of articles from the Special Issue published online in the open access journal *Processes* (ISSN 2227-9717) (available at: https://www.mdpi.com/journal/processes/special_issues/hydrogen_production).

For citation purposes, cite each article independently as indicated on the article page online and as indicated below:

LastName, A.A.; LastName, B.B.; LastName, C.C. Article Title. *Journal Name* **Year**, *Article Number*, Page Range.

ISBN 978-3-03943-667-5 (Hbk)
ISBN 978-3-03943-668-2 (PDF)

© 2020 by the authors. Articles in this book are Open Access and distributed under the Creative Commons Attribution (CC BY) license, which allows users to download, copy and build upon published articles, as long as the author and publisher are properly credited, which ensures maximum dissemination and a wider impact of our publications.

The book as a whole is distributed by MDPI under the terms and conditions of the Creative Commons license CC BY-NC-ND.

Contents

About the Editors vii

Suttichai Assabumrungrat, Suwimol Wongsakulphasatch, Pattaraporn Lohsoontorn Kim and Alírio E. Rodrigues
Special Issue on "Hydrogen Production Technologies"
Reprinted from: *Processes* **2020**, *8*, 1268, doi:10.3390/pr8101268 1

Watcharapong Khaodee, Tara Jiwanuruk, Khunnawat Ountaksinkul, Sumittra Charojrochkul, Jarruwat Charoensuk, Suwimol Wongsakulphasatch and Suttichai Assabumrungrat
Compact Heat Integrated Reactor System of Steam Reformer, Shift Reactor and Combustor for Hydrogen Production from Ethanol
Reprinted from: *Processes* **2020**, *8*, 708, doi:10.3390/pr8060708 5

R. Visvanichkul, S. Peng-Ont, W. Ngampuengpis, N. Sirimungkalakul, P. Puengjinda, T. Jiwanuruk, T. Sornchamni and P. Kim-Lohsoontorn
Effect of CuO as Sintering Additive in Scandium Cerium and Gadolinium-Doped Zirconia-Based Solid Oxide Electrolysis Cell for Steam Electrolysis
Reprinted from: *Processes* **2019**, *7*, 868, doi:10.3390/pr7120868 25

Nonchanok Ngoenthong, Matthew Hartley, Thana Sornchamni, Nuchanart Siri-nguan, Navadol Laosiripojana and Unalome Wetwatana Hartley
Comparison of Packed-Bed and Micro-Channel Reactors for Hydrogen Production via Thermochemical Cycles of Water Splitting in the Presence of Ceria-Based Catalysts
Reprinted from: *Processes* **2019**, *7*, 767, doi:10.3390/pr7100767 35

Li Xu, Ying Wang, Syed Ahsan Ali Shah, Hashim Zameer, Yasir Ahmed Solangi, Gordhan Das Walasai and Zafar Ali Siyal
Economic Viability and Environmental Efficiency Analysis of Hydrogen Production Processes for the Decarbonization of Energy Systems
Reprinted from: *Processes* **2019**, *7*, 494, doi:10.3390/pr7080494 47

Supanida Chimpae, Suwimol Wongsakulphasatch, Supawat Vivanpatarakij, Thongchai Glinrun, Fasai Wiwatwongwana, Weerakanya Maneeprakorn and Suttichai Assabumrungrat
Syngas Production from Combined Steam Gasification of Biochar and a Sorption-Enhanced Water–Gas Shift Reaction with the Utilization of CO_2
Reprinted from: *Processes* **2019**, *7*, 349, doi:10.3390/pr7060349 71

Bo Chen, Tao Yang, Wu Xiao and Aazad khan Nizamani
Conceptual Design of Pyrolytic Oil Upgrading Process Enhanced by Membrane-Integrated Hydrogen Production System
Reprinted from: *Processes* **2019**, *7*, 284, doi:10.3390/pr7050284 87

Diksha Kapoor, Amandeep Singh Oberoi and Parag Nijhawan
Hydrogen Production and Subsequent Adsorption/Desorption Process within a Modified Unitized Regenerative Fuel Cell
Reprinted from: *Processes* **2019**, *7*, 238, doi:10.3390/pr7040238 105

William J. F. Gannon, Daniel R. Jones and Charles W. Dunnill
Enhanced Lifetime Cathode for Alkaline Electrolysis Using Standard Commercial Titanium Nitride Coatings
Reprinted from: *Processes* **2019**, *7*, 112, doi:10.3390/pr7020112 123

Elvira Tapia, Aurelio González-Pardo, Alfredo Iranzo, Manuel Romero, José González-Aguilar, Alfonso Vidal, Mariana Martín-Betancourt and Felipe Rosa
Multi-Tubular Reactor for Hydrogen Production: CFD Thermal Design and Experimental Testing
Reprinted from: *Processes* **2019**, *7*, 31, doi:10.3390/pr7010031 . . . . **135**

Asad A. Zaidi, Ruizhe Feng, Adil Malik, Sohaib Z. Khan, Yue Shi, Asad J. Bhutta and Ahmer H. Shah
Combining Microwave Pretreatment with Iron Oxide Nanoparticles Enhanced Biogas and Hydrogen Yield from Green Algae
Reprinted from: *Processes* **2019**, *7*, 24, doi:10.3390/pr7010024 . . . . **151**

About the Editors

Suttichai Assabumrungrat is Professor of Chemical Engineering at Faculty of Engineering, Chulalongkorn University. His research interests are in process intensification with particular focus on multifunctional reactors. He has published about 300 peer-reviewed journal and proceedings articles and book chapters. He is now working on several projects, for example, involving hydrogen production technologies, biodiesel production, biorefineries, and CO_2 capture and utilization.

Suwimol Wongsakulphasatch is Associate Professor of Chemical Engineering at Faculty of Engineering, King Mongkut's University of Technology North Bangkok. Her research areas of interest are in H_2 production via thermochemical processes integrated with CO_2 capture by porous solids.

Pattaraporn Lohsoontorn Kim is Assistant Professor in Chemical Engineering at Faculty of Engineering, Chulalongkorn University. Her current research interests are focused on electrochemical devices for energy applications, with particular attention given to fuel cells and electrolyzers. Her research interests also include thermochemical and electrochemical CO_2 conversion to higher value products such as methanol, other chemicals, and carbon products.

Alírio E. Rodrigues is Emeritus Professor in Chemical Engineering at Faculty of Engineering, University of Porto. His research interests are in the areas of cyclic adsorption/reaction processes for process intensification, perfume engineering and microencapsulation, and lignin valorization. He has published about 700 peer-reviewed journal articles, several books and patents. He is now working on carbon capture and utilization, power-to-gas processes, gas phase simulated moving beds for olefin/paraffin separation, supercritical simulated moving bed reactors, biorefinery processes, and trails of perfumes.

Editorial

Special Issue on "Hydrogen Production Technologies"

Suttichai Assabumrungrat [1,2,*], Suwimol Wongsakulphasatch [3], Pattaraporn Lohsoontorn Kim [1,2] and Alírio E. Rodrigues [4]

[1] Center of Excellence on Catalysis and Catalytic Reaction Engineering, Department of Chemical Engineering, Faculty of Engineering, Chulalongkorn University, Bangkok 10330, Thailand; Pattaraporn.K@chula.ac.th
[2] Bio-Circular-Green-economy Technology & Engineering Center, BCGeTEC, Department of Chemical Engineering, Faculty of Engineering, Chulalongkorn University, Bangkok 10330, Thailand
[3] Department of Chemical Engineering, Faculty of Engineering, King Mongkut's University of Technology North Bangkok, Bangkok 10800, Thailand; suwimol.w@eng.kmutnb.ac.th
[4] Laboratory of Separation and Reaction Engineering–Laboratory of Catalysis and Materials, Departamento de Engenharia Química, Faculdade de Engenharia da Universidade do Porto, 4200-465 Porto, Portugal; arodrig@fe.up.pt
* Correspondence: suttichai.a@chula.ac.th

Received: 27 September 2020; Accepted: 3 October 2020; Published: 9 October 2020

According to energy crisis and environmental concerns, hydrogen has been driven to become one of the most promising alternative energy carriers for power generation and high valued chemical products. To meet the requirements of global energy demand, which continuously increase each year, efficient technologies to produce hydrogen are therefore necessary. This Special Issue on "Hydrogen Production Technologies" covers outstanding researches and technologies to produce hydrogen of which their objectives are to improve process performances. Both theoretical and experimental investigations were conducted for the investigation of parametric effects in terms of technical and/or economical aspects for different routes of hydrogen production technologies, including thermochemical, electrochemical, and biological. In addition, techniques used to storage and utilize hydrogen were also demonstrated.

Steam electrolysis reaction is a technique used to produce hydrogen through solid oxide electrolysis cells (SOECs). Visvanichkul et al. [1] studied the effect of CuO addition into $Sc_{0.1}Ce_{0.05}Gd_{0.05}Zr_{0.89}O_2$ (SCGZ) electrolyte as a sintering additive on phase formation, cell densification, and electrical performance at elevated temperature. The results showed significant effect on the sinter ability of SCGZ. With the addition of 0.5 wt% CuO phase transformation and impurity were not observed. However, the sintering ability of the electrolyte achieved 95% relative density with a large grain size at 1573 K. Electrochemical performance evaluated at the operating temperature ranging from 873 K to 1173 K under steam to hydrogen ratio at 70:30 showed activation energy of conduction (E_a) of the SCGZ with CuO of 74.93 kJ mol^{-1} compared to that without Cu of 72.34 kJ mol^{-1}. Another work presented by Gannon et al. [2] was conducted in improving performance of electrode for water splitting under room temperature. Titanium nitride coating 316 grade stainless-steel electrode was found to be able to extend the electrode lifetime to over 2000 cycles lasting 5.5 days and was observed to outperform the uncoated material by 250 mV.

An alternative route for hydrogen production is from the conversion of solar energy. Tapia et al. [3] investigated the use of multi-tubular solar reactors for hydrogen production through thermochemical cycle using CFD modelling and simulations to design the reactor for a pilot plant in the Plataforma Solar de Almería (PSA). The developed CFD model showed its validated results with the experimental data having a temperature error ranging from 1% to around 10%, depending upon the location inside the reactor. The thermal balance solved by the CFD model revealed a 7.9% thermal efficiency of the reactor, and ca. 90% of the ferrite domain could achieve the required process temperature of 900 °C. Treatment of reactants before producing hydrogen is another technique that helps to enhance process efficiency. Zaidi et al. [4] studied the effect of using microwave (MW) and Fe_3O_4 nanoparticles (NPs) to improve biodegradability of green algae, yielded biogas—a source of hydrogen production. Their results showed both yields of biogas and hydrogen could be improved when compared to the individual ones. The biogas amount of 328 mL and 51.5% v/v hydrogen were produced by MW pretreatment + Fe_3O_4 NPs.

Integrated techniques to improve hydrogen production performances have also been investigated. Ngoenthong et al. [5] developed a catalyst for hydrogen production from a two-step thermochemical cycle of water-splitting, applied with two different reactor types, packed-bed and micro-channel reactors. Ceria-zirconia ($Ce_{0.75}Zr_{0.25}O_2$) was found to offer better catalytic activity than fluorite-structure ceria (CeO_2), which was suggested to be due to higher oxygen storage capacity. The micro-channel reactor showed 16 times higher H_2 productivity than the packed-bed reactor at the same operating temperature of 700 °C. The better performance of the micro-channel reactor was considered as a result of high surface-to-volume ratio of the reactor, facilitating accessibility of the reactant molecules to react on the catalyst surface. Chimpae et al. [6] evaluated performance of a combined gasification and a sorption-enhanced water–gas shift reaction (SEWGS) for synthesis gas production using mangrove-derived biochar as a feedstock. Multifunctional material was applied in this integrated process and the effects of biochar gasification temperature, pattern of combined gasification and SEWGS, amount of co-fed steam and CO_2 as gasifying agent, and SEWGS temperature were studied. The studies revealed that the hybrid process could produce greater amount of H_2 with a lower amount of CO_2 emissions when compared with separated sorbent/catalyst material. Syngas production was found to depend upon the composition of gasifying agent and SEWGS temperature. An integrated steam methane reforming-hydrotreating (SMR-HT) pyrolytic oil upgrading process enhanced by membrane gas separation system was proposed by Chen et al. [7]. Process design and process optimization were developed through simulation framework of commercial software Aspen HYSIS along with the developed self-defined extensions for Aspen HYSYS. The results revealed that the proposed process could provide 63.7% conversion with 2.0 wt% hydrogen consumption and 70% higher net profit could be obtained when compared with the conventional process. Khaodee et al. [8] proposed systems of compact heat integrated reactor system (CHIRS) of a steam reformer, a water gas shift reactor, and a combustor for stationary hydrogen production from ethanol. Their performances were simulated using COMSOL Multiphysics software.

As there are a number of different techniques that could be used to produce hydrogen, we therefore need to consider a selection of technologies for its production. One tool that could be used to assist decision making is data analysis. Xu et al. [9] developed a framework includes slack-based data envelopment analysis (DEA), with fuzzy analytical hierarchy process (FAHP), and fuzzy technique for order of preference by similarity to ideal solution (FTOPSIS), to prioritize hydrogen production in Pakistan. Five criteria, including capital cost, feedstock cost, O&M cost, hydrogen production, and CO_2 emission were taken into consideration. The results showed that wind electrolysis, PV electrolysis, and biomass gasification offered fully efficient and were recommended as sustainable selections for production of hydrogen in Pakistan.

High production of hydrogen demand leads also to high demand of efficient hydrogen storage system. Kapoor et al. [10] developed electrochemical hydrogen storage by integrating a solid multi-walled carbon nanotube (MWCNT) electrode in a modified unitized regenerative fuel cell (URFC). A method to fabricate solid electrode from MWCNT powder and egg white as an organic binder was investigated. The results showed that the developed porous MWCNT electrode had electrochemical hydrogen storage capacity of 2.47wt%, comparable with commercially available AB_5-based hydrogen storage canisters.

All the above papers show high-quality research articles on various innovative hydrogen production related technologies. The works and topics address current status and future challenges in unit scale and overall process performances. Under the high demand of renewable and sustainable energy at present, we believe that these articles would find beneficial to a wide interest of readers.

We thank Managing Editor, Ms. Jamie Li, all *Processes* staff, and all contributors, for enthusiastic and kindly support of this Special Issue.

Suttichai Assabumrungrat
Suwimol Wongsakulphasatch
Pattaraporn Lohsoontorn Kim
Alírio E. Rodrigues
Guest Editors

Funding: This research received no external funding.

Conflicts of Interest: The authors declare no conflict of interest.

References

1. Visvanichkul, R.; Peng-Ont, S.; Ngampuengpis, W.; Sirimungkalakul, N.; Puengjinda, P.; Jiwanuruk, T.; Sornchamni, T.; Kim-Lohsoontorn, P. Effect of CuO as Sintering Additive in Scandium Cerium and Gadolinium-Doped Zirconia-Based Solid Oxide Electrolysis Cell for Steam Electrolysis. *Processes* **2019**, *7*, 868. [CrossRef]
2. Gannon, W.; Jones, D.; Dunnill, C. Enhanced Lifetime Cathode for Alkaline Electrolysis Using Standard Commercial Titanium Nitride Coatings. *Processes* **2019**, *7*, 112. [CrossRef]
3. Tapia, E.; González-Pardo, A.; Iranzo, A.; Romero, M.; González-Aguilar, J.; Vidal, A.; Martín-Betancourt, M.; Rosa, F. Multi-Tubular Reactor for Hydrogen Production: CFD Thermal Design and Experimental Testing. *Processes* **2019**, *7*, 31. [CrossRef]
4. Zaidi, A.; Feng, R.; Malik, A.; Khan, S.; Shi, Y.; Bhutta, A.; Shah, A. Combining Microwave Pretreatment with Iron Oxide Nanoparticles Enhanced Biogas and Hydrogen Yield from Green Algae. *Processes* **2019**, *7*, 24. [CrossRef]
5. Ngoenthong, N.; Hartley, M.; Sornchamni, T.; Siri-nguan, N.; Laosiripojana, N.; Hartley, U. Comparison of Packed-Bed and Micro-Channel Reactors for Hydrogen Production via Thermochemical Cycles of Water Splitting in the Presence of Ceria-Based Catalysts. *Processes* **2019**, *7*, 767. [CrossRef]
6. Chimpae, S.; Wongsakulphasatch, S.; Vivanpatarakij, S.; Glinrun, T.; Wiwatwongwana, F.; Maneeprakorn, W.; Assabumrungrat, S. Syngas Production from Combined Steam Gasification of Biochar and a Sorption-Enhanced Water–Gas Shift Reaction with the Utilization of CO_2. *Processes* **2019**, *7*, 349. [CrossRef]
7. Chen, B.; Yang, T.; Xiao, W.; Nizamani, A. Conceptual Design of Pyrolytic Oil Upgrading Process Enhanced by Membrane-Integrated Hydrogen Production System. *Processes* **2019**, *7*, 284. [CrossRef]
8. Khaodee, W.; Jiwanuruk, T.; Ountaksinkul, K.; Charojrochkul, S.; Charoensuk, J.; Wongsakulphasatch, S.; Assabumrungrat, S. Compact Heat Integrated Reactor System of Steam Reformer, Shift Reactor and Combustor for Hydrogen Production from Ethanol. *Processes* **2020**, *8*, 708. [CrossRef]
9. Xu, L.; Wang, Y.; Shah, S.; Zameer, H.; Solangi, Y.; Walasai, G.; Siyal, Z. Economic Viability and Environmental Efficiency Analysis of Hydrogen Production Processes for the Decarbonization of Energy Systems. *Processes* **2019**, *7*, 494. [CrossRef]
10. Kapoor, D.; Oberoi, A.; Nijhawan, P. Hydrogen Production and Subsequent Adsorption/Desorption Process within a Modified Unitized Regenerative Fuel Cell. *Processes* **2019**, *7*, 238. [CrossRef]

Publisher's Note: MDPI stays neutral with regard to jurisdictional claims in published maps and institutional affiliations.

© 2020 by the authors. Licensee MDPI, Basel, Switzerland. This article is an open access article distributed under the terms and conditions of the Creative Commons Attribution (CC BY) license (http://creativecommons.org/licenses/by/4.0/).

Article

Compact Heat Integrated Reactor System of Steam Reformer, Shift Reactor and Combustor for Hydrogen Production from Ethanol

Watcharapong Khaodee [1,2,*], Tara Jiwanuruk [3], Khunnawat Ountaksinkul [3], Sumittra Charojrochkul [4], Jarruwat Charoensuk [5], Suwimol Wongsakulphasatch [6] and Suttichai Assabumrungrat [3]

[1] Department of Chemical Engineering, Mahanakorn University of Technology, Nong Chok, Bangkok 10530, Thailand
[2] Chemical Engineering Program, Department of Industrial Engineering, Faculty of Engineering, Naresuan University, Phitsanulok 65000, Thailand
[3] Department of Chemical Engineering, Center of Excellence in Catalysis and Catalytic Reaction Engineering, Faculty of Engineering, Chulalongkorn University, Bangkok 10330, Thailand; tara_pkwb@hotmail.com (T.J.); khunnawat_de@hotmail.com (K.O.); suttichai.a@chula.ac.th (S.A.)
[4] National Metal and Materials Technology Center (MTEC), Pathumthani 12120, Thailand; sumittrc@mtec.or.th
[5] Mechanical Engineering Department, Faculty of Engineering, King Mongkut's Institute of Technology Ladkrabang, Bangkok 10520, Thailand; kcjarruw@kmitl.ac.th
[6] Department of Chemical Engineering, Faculty of Engineering, King Mongkut's University of Technology North Bangkok, Bangkok 10800, Thailand; suwimol.w@eng.kmutnb.ac.th
* Correspondence: watcharapongk@nu.ac.th or kwatchar@mut.ac.th; Tel.: +66-5596-4204

Received: 18 April 2019; Accepted: 16 June 2020; Published: 19 June 2020

Abstract: A compact heat integrated reactor system (CHIRS) of a steam reformer, a water gas shift reactor, and a combustor were designed for stationary hydrogen production from ethanol. Different reactor integration concepts were firstly studied using Aspen Plus. The sequential steam reformer and shift reactor (SRSR) was considered as a conventional system. The efficiency of the SRSR could be improved by more than 12% by splitting water addition to the shift reactor (SRSR-WS). Two compact heat integrated reactor systems (CHIRS) were proposed and simulated by using COMSOL Multiphysics software. Although the overall efficiency of the CHIRS was quite a bit lower than the SRSR-WS, the compact systems were properly designed for portable use. CHIRS (I) design, combining the reactors in a radial direction, was large in reactor volume and provided poor temperature control. As a result, the ethanol steam reforming and water gas shift reactions were suppressed, leading to lower hydrogen selectivity. On the other hand, CHIRS (II) design, combining the process in a vertical direction, provided better temperature control. The reactions performed efficiently, resulting in higher hydrogen selectivity. Therefore, the high performance CHIRS (II) design is recommended as a suitable stationary system for hydrogen production from ethanol.

Keywords: compact reactor; ethanol steam reforming; water gas shift; hydrogen production

---

1. Introduction

Hydrogen has been used widely in many industrial processes such as petroleum, petrochemical, steel and food. Nowadays, hydrogen, which is also considered as a clean fuel, has recently been used in vehicular systems to reduce fossil fuel usage [1,2]. Therefore, hydrogen utilization demand has dramatically increased, leading to insufficient hydrogen supply with restricted hydrogen sources. In conventional processes, hydrogen is produced from steam reforming of natural gas (NG) or liquefied petroleum gas (LPG). The process emits gaseous carbon dioxide and causes environmental problems. Alternative green fuels such as biogas, ethanol and bio-oil have been suggested for hydrogen production [3–9]. Ethanol, a harmless liquid at room temperature, has potential to be a good candidate for steam reforming [10–17], since it can be produced from agricultural products and bio-waste fermentation.

Ethanol steam reforming is a highly endothermic reaction, and it produces various by-products, such as methane and acetaldehyde [18,19]. The reaction normally occurs at high temperatures, beyond 973 K, to reduce the by-products. An external heat source is required to maintain a high temperature for the reforming reaction. However, a reverse water gas shift strongly occurs at high temperature. The reaction produces carbon monoxide and decreases hydrogen production. Thus, a water gas shift reactor, which operates at lower temperatures, is necessary to shift the reaction equilibrium and to increase the hydrogen production rate [20–24]. A compact reactor system consisting of combustor, reformer and shift reactor is proposed in this study for hydrogen production from ethanol.

For hydrogen production, multifunctional reactors, which combine a combustor within a reformer, have been studied extensively. For these reactors, heat is normally generated from combustion and transfers to the reformer side through the reactor's wall [25–28]. A microreactor, for instance, consisting of two parallel channels for methanol combustion and methanol steam reforming was studied by Andisheh Tadbir and Akbari [25]. An assembly of 1540 small reactor sets occupying a total volume of about 91 cm^3 can produce enough hydrogen for operating a typical 30-W PEM fuel cell. A reformer that integrated the steam reforming reaction and catalytic combustion in a reactor was also investigated by Grote et al. [26]. Experiments and simulations were employed and the model was successfully validated with experimental data of 4 kW, 6 kW and 10 kW reformers. As reported in another study, a metallic monolith catalyst for methane catalytic combustion and methane dry reforming was examined by Yin et al. [27]. Methane conversion in dry reforming reached 93.6% with 81.9% of heat efficiency. For water gas shift reactor integration, a compact steam reformer was investigated numerically by Seo et al. [29]. Methane was converted to syngas in a steam reforming section and then flowed to a water gas shift section. In the product stream, methane conversion and CO concentration were 87% and 0.45%, respectively. Furthermore, Hayer et al. employed the integrated micro packed bed reactor heat exchanger (IMPBRHE) for the synthesis of dimethyl ether [30]. This work presented a comparison between the temperature profiles along the length of IMPBRHE and that of the fixed bed reactor under the same operating conditions, investigated via COMSOL Multiphysics. Their results showed that the temperature gradients in the microchannel reactor were steeper than those in the lab-scale fixed bed reactor. It could be concluded that the microchannel reactor offered high heat transfer due to its high surface area-to-volume ratio. All of these factors indicate that the multifunctional reactor combined with the compact system has a high possibility for hydrogen production from ethanol, giving two benefits as follows: (1) heat integration to optimize energy consumption and (2) good mass and heat transfer owing to high surface area to volume ratio.

This study aimed to design a compact reactor system consisting of a combustor, a steam reformer and a shift reactor for stationary hydrogen production from ethanol. Regarding the step of process concept development, different processes integration concepts including typical sequential steam reformer and shift reactor (SRSR), SRSR with energy management by water splitting (SRSR-WS) and a compact heat integrated reactor system (CHIRS) were preliminarily examined via Aspen Plus software. The highest level of process concept development, CHIRS, was further studied in detail via COMSOL Multiphysics software. The Aspen Plus software was used to determine the suitable concept from the three integration concepts mentioned above, whereas the COMSOL Multiphysics software was applied to investigate the transport phenomena inside the reactors and finally to determine the proper configuration of the suitable case considered by Aspen Plus software.

2. Modeling and Simulation

2.1. Description of Reformer Concept Development

As illustrated in Figure 1, there are three steps of ethanol steam reformer concept development considered in this work, i.e., typical sequential steam reformer and shift reactor (SRSR), SRSR with energy management by water splitting (SRSR-WS) and compact heat integrated reactor system (CHIRS). For the conventional one, SRSR, ethanol and water are fed to the reformer at the desired temperature and pressure and the product stream flows to the shift reactor at the same operating condition as shown in Figure 1a. However, to operate the shift reactor efficiently, it should be carried out at low temperature to achieve a higher hydrogen production. Therefore, for the second level of process concept development, SRSR-WS, additional water is used to mix with the product stream of the reformer to quench to the desired temperature of the shift reactor. This stream is then fed to the adiabatic shift reactor (Figure 1b). However, for the first two concepts, the heat management in each process has not been considered. For example, an ethanol steam reformer typically requires heat from the external heat source and heat from the product stream at high temperature to be recovered. Hence, the heat management is intentionally included in the final step of process concept development, CHIRS. As displayed in Figure 1c, the heat requirement for the process is supplied from a combustor, which uses methane as a fuel. Two heat exchangers are installed to preheat the reactant. Moreover, the reformed gas from the ethanol steam reformer at high temperature can be reduced to the suitable temperature for the shift reactor by diverting heat to the reactant via heat exchanger I. To reduce heat loss at the outlet and improve the process efficiency, the temperature of the combusted gas after exchanging heat with the reactant (Heat exchanger II) is properly limited at 523 K.

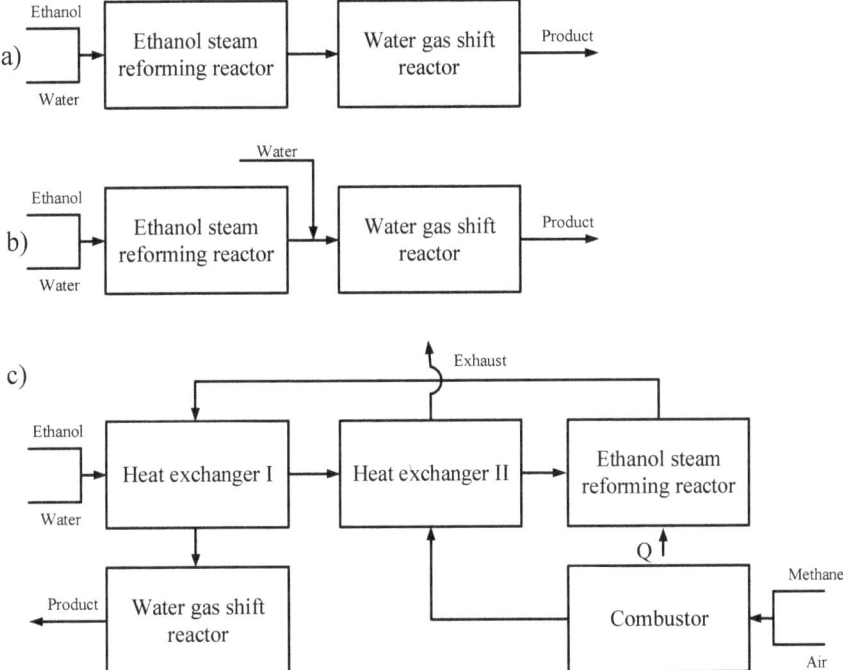

Figure 1. Different ethanol steam reformer concepts: (**a**) Typical sequential reformer and shift reactor (SRSR), (**b**) SRSR with energy management by water splitting (SRSR-WS), and (**c**) Compact heat integrated reactor systems (CHIRS).

To evaluate these concepts, a simulation via Aspen Plus software was selected. Aspen Plus software is widely used for process simulations in chemical industries. The program contains standard and ideal unit operations such as reactor and heat exchanger models. For steam reforming processes, reactors including an ethanol steam reformer, a shift reactor, a combustor and a heat exchanger are generally conducted in the simulation. To simplify the simulations, a thermodynamic equilibrium reactor, RGibbs reactor, was assumed. In the RGibbs reactor model, Gibbs free energy minimization is performed to determine the product compositions. For an ethanol steam reforming reaction, possible products including hydrogen (H_2), carbon monoxide (CO), carbon dioxide (CO_2), methane (CH_4), acetaldehyde (CH_3CHO), acetone (C_3H_6O), dimethyl ether (C_2H_6O), ethane (C_2H_6), ethylene (C_2H_4) and coke (C) were specified [18,19,31].

In the simulations, ethanol reactant was fed at 1 kmol/h at standard temperature and pressure. The operating condition for the ethanol steam reforming reaction was estimated to find an appropriate range of temperature, pressure and steam to ethanol ratio as discussed later in Section 3.1.1. A proper condition was employed in the reformer concept investigation. Efficiency as defined in Equation (1) is used as an indicator in this study.

$$\text{Efficiency (in \%)} = \frac{\dot{n}_{H_2} \cdot \Delta H_{c,H_2}}{\dot{n}_{Ethanol} \cdot \Delta H_{c,Ethanol} + \text{input energy}} \times 100 \qquad (1)$$

2.2. CHIRS Designs in Detail

This compact reactor system, combining combustor, reformer, shift reactor and two heat exchangers within a structure, was designed to be the same as the combined reformer with heat exchanger network concept. The process was developed and named as a compact heat integrated reactor system (CHIRS). For the first CHIRS design (CHIRS (I)) as illustrated in Figure 2, the ethanol steam reformer was placed inside the combustion chamber. The reformer received heat directly from the combustion through the reactor's wall, which performed as a heat exchanger. The reformed gas from the reformer was fed to the shift reactor located in the air preheat chamber. The reformed gas was quenched by air and the water gas shift reaction was shifted forward, leading to an increase in hydrogen production. For the CHIRS (I) design, the sections were integrated in the radial direction. The combustion chamber was enveloped by an air preheat chamber as shown in Figure 2a. However, there was another interesting design designated as CHIRS (II), which combined the processes in the vertical direction as shown in Figure 3. For the second design, an air gap insulator was set between the reformer and the shift reactor.

Owing to the study of CHIRS in detail, three-dimensional computational fluid dynamic (CFD) simulation was employed to examine the process performance of CHIRS design using COMSOL Multiphysics software. The gray area presented in Figures 2a and 3a was set as the calculation domain for CHIRS (I) and CHIRS (II), respectively. Tetrahedral mesh was created to cover the structure. The mesh size was specified as extremely fine inside the ethanol steam reforming and water gas shift reactors due to the presence of reaction in these domains obtaining a high gradient in concentration and temperature profiles. Total mesh number of CHIRS (I), 4.02×10^5 elements, was higher than that of CHIRS (II), 2.40×10^5 elements, because the former was larger in size than the latter.

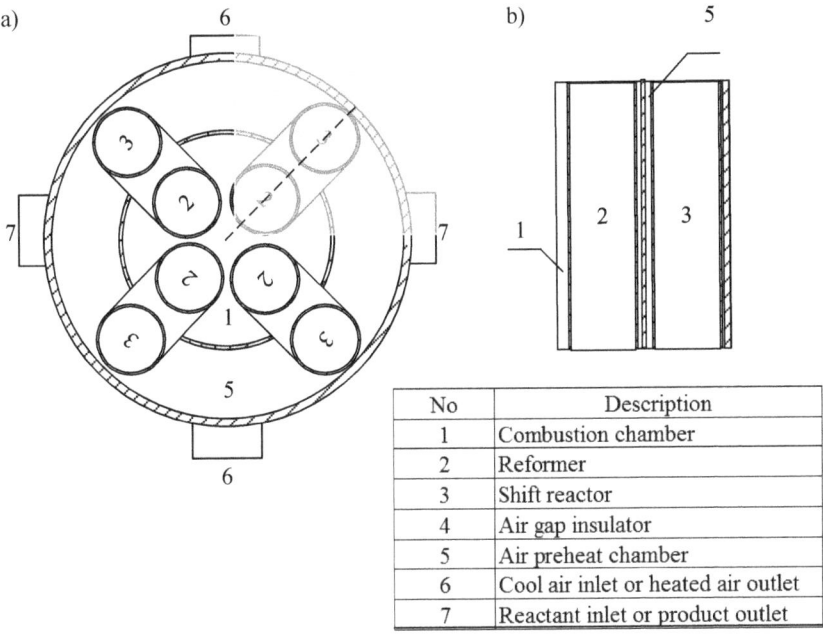

Figure 2. Configuration of CHIRS (I) shown in (a) top view and (b) cross sectional view (dotted line).

No	Description
1	Combustion chamber
2	Reformer
3	Shift reactor
4	Air gap insulator
5	Air preheat chamber
6	Cool air inlet or heated air outlet
7	Reactant inlet or product outlet

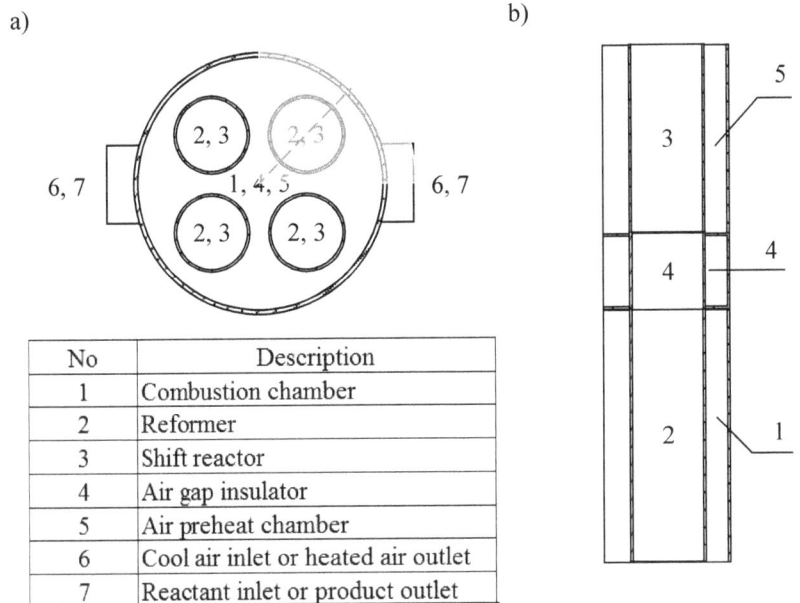

Figure 3. Configuration of CHIRS (II), shown in (a) top view and (b) cross sectional view (dotted line).

No	Description
1	Combustion chamber
2	Reformer
3	Shift reactor
4	Air gap insulator
5	Air preheat chamber
6	Cool air inlet or heated air outlet
7	Reactant inlet or product outlet

Inside the module of COMSOL Multiphysics, several governing equations are taken into account. The steady state governing equations, i.e., mass, momentum, energy and chemical species conservation equations, which can be written in Equations (2)–(5), respectively, were simultaneously considered.

$$\rho(\nabla \cdot \vec{v}) = 0 \tag{2}$$

$$\rho(\vec{v} \cdot \nabla \vec{v}) = -\nabla p + \nabla \cdot \left[\mu\left(\nabla \vec{v} + \nabla \vec{v}^T\right)\right] + \rho \vec{g} \tag{3}$$

$$\rho \nabla \cdot (C_p \vec{v} T) = \nabla \cdot (k \nabla T) + \sum_j (\Delta H_j r_j) \tag{4}$$

$$\rho \nabla \cdot (\vec{v} \omega_i) = \rho \nabla \cdot (D_{i,eff} \nabla \omega_i) + \sum_i (r_j MW) \tag{5}$$

The gravity term in Equation (3) was neglected. The related reactions, which were ethanol steam reforming and water gas shift, were computed using kinetic models. For the ethanol steam reforming reaction, the reactions are divided into Equations (R1)–(R4).

$$C_2H_5OH \rightarrow CH_3CHO + H_2 \tag{R1}$$

$$C_2H_5OH \rightarrow CH_4 + CO + H_2 \tag{R2}$$

$$CO + H_2O \leftrightarrow CO_2 + H_2 \tag{R3}$$

$$CH_3CHO + 3H_2O \leftrightarrow 2CO_2 + 5H_2 \tag{R4}$$

Kinetic models of these reactions over a Co_3O_4–ZnO catalyst were adopted from Uriz et al. as listed in Equations (6)–(9) [32].

$$r_{R1} = 2.1 \times 10^4 \exp\left(\frac{-70(kJ/mol)}{Rg} \cdot \left(\frac{1}{T} - \frac{1}{773}\right)\right) \times P_{C_2H_5OH} \tag{6}$$

$$r_{R2} = 2.0 \times 10^3 \exp\left(\frac{-130(kJ/mol)}{Rg} \cdot \left(\frac{1}{T} - \frac{1}{773}\right)\right) \times P_{C_2H_5OH} \tag{7}$$

$$r_{R3} = 1.9 \times 10^4 \exp\left(\frac{-70(kJ/mol)}{Rg} \cdot \left(\frac{1}{T} - \frac{1}{773}\right)\right) \times \left(P_{CO}P_{H_2O} - \frac{P_{CO_2}P_{H_2}}{K_{WGS}}\right) \tag{8}$$

$$r_{R4} = 2.0 \times 10^5 \exp\left(\frac{-98(kJ/mol)}{Rg} \cdot \left(\frac{1}{T} - \frac{1}{773}\right)\right) \times P_{CH_3CHO}P_{H_2O}^3 \tag{9}$$

where P_i is partial pressure of component i in bar and K_{WGS} is defined as shown in Equation (10).

$$K_{WGS} = \exp\left(\frac{4577.8}{T} - 4.33\right) \tag{10}$$

In the shift reactor, the $Cu/ZnO/Al_2O_3$ catalyst has been typically used for the water gas shift reaction (Equation (R3)). The kinetic model was proposed by Amadeo and Laborde as listed in Equation (11) [22].

$$r_{WGS} = \frac{0.92e^{(-454.3/T)}P_{CO}P_{H_2O}(1 - P_{CO_2}P_{H_2}/P_{CO}P_{H_2O}K_{WGS})}{\left(1 + 2.2e^{(101.5/T)}P_{CO} + 0.4e^{(158.3/T)}P_{H_2O} + 0.0047e^{(2737.9/T)}P_{CO_2} + 0.05e^{(1596.1/T)}P_{H_2}\right)^2} \tag{11}$$

Fluid properties were simplified and assumed as steam and air for the reforming stream and combusted gas, respectively. The reactor structure was stainless steel. The porous media was considered as alumina according to the general catalyst support material.

3. Results and Discussion

3.1. Preliminary Study of Reformer via Aspen Plus

3.1.1. Effect of Operating Conditions on Reaction Performance

Operating parameters including temperature, pressure and steam to ethanol ratio were determined to find a suitable operation range that provided high hydrogen production without coke formation in the process. To study the effects of temperature and pressure, water was firstly fed at 3 kmol/h in the conventional SRSR reactor according to the stoichiometry of the ethanol steam reforming reaction.

When considering the atmospheric pressure (1 atm), the effect of operating temperature in the range of 400–1300 K on product distribution was reported as shown in Figure 4a. Methane and syngas were mainly produced in this reforming temperature range. Ethanol was completely converted to intermediate gas while by-products including acetaldehyde, acetone, dimethyl ether, ethane and ethylene were absent in the product stream due to the non-thermodynamic stability of these components [19,31]. Methane steam reforming and water gas shift were main reactions in this reformer. Hydrogen production increased with an increase in reforming temperature and reached the optimum conversion at 1023 K. Below 1023 K, hydrogen production was increased as a result of methane steam reforming, which was a major reaction, whereas the reduction in hydrogen production at a higher temperature than 1023 K occurred because hydrogen was reasonably consumed by the reverse water gas shift reaction. Eventually, the ethanol steam reforming for hydrogen production was appropriately carried out at a moderate temperature of 1023 K.

As shown in Figure 4b, the effect of pressure in the range of 1–5 atm at the proper temperature, 1023 K, was further investigated. As the operating pressure increased, the methane steam reforming was suppressed, resulting in the reduction in hydrogen production with increasing methane composition in the reformed gas. The reaction equilibrium shifted backward as the pressure increased according to mole expansion of the steam reforming reaction. Therefore, the hydrogen production from ethanol was preferentially operated at low operating pressure, especially 1 atm, due to simple design and operation.

Coke formation, which causes catalyst deactivation and limits the operation time, is an important indicator for operating condition selection. According to Montero et al. [33], acetaldehyde, ethylene, and non-reacted ethanol are main precursors for coke formation on the metal sites at low space-time. At high space-time, due to a change in the coke mechanism, the CH_4 and CO become the main precursors leading to a filamentous and partially graphitic coke. The increases of temperature and ethanol to steam ratio along with a significantly prolonged reaction lead to coke formation. The catalyst deactivation is attenuated by reducing the concentration of coke precursors and increasing coke gasification, especially at high temperature. Therefore, this study considered the coke formation at the reforming pressure of 1 atm with various reforming temperatures and steam to ethanol ratios as presented in Figure 5. Coke formation decreased with increasing operating temperature and steam to ethanol ratio. When the steam to ethanol ratio was below 3, coke strongly appeared over the reforming temperature range of 400–1300 K. Beyond the reforming temperature of 523 K and a steam to ethanol ratio of 3, coke formation then became negligible. Thus, an operating condition at 523 K and steam to ethanol ratio of 3 was the lowest boundary for the ethanol steam reforming process without any coke formation.

An appropriate operating condition for ethanol steam reforming was at the reforming temperature of 573–1073 K, steam to ethanol ratio of 3–5 and reforming pressure of 1 atm. This condition provided high hydrogen production without coke formation and was further employed in the reformer concept study.

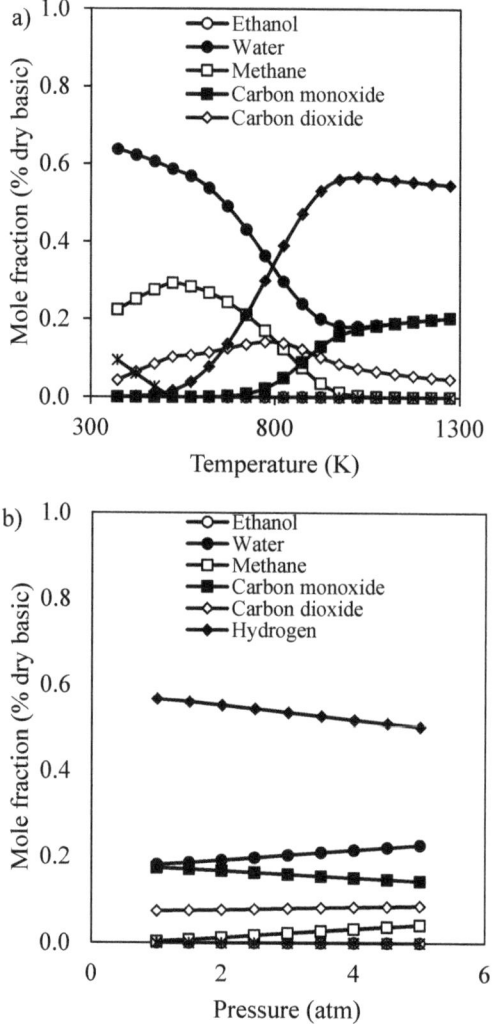

Figure 4. Mole fraction of product stream at a steam to ethanol ratio of 3 for (**a**) effect of temperature ($P = 1$ atm), and (**b**) effect of pressure ($T = 1023$ K).

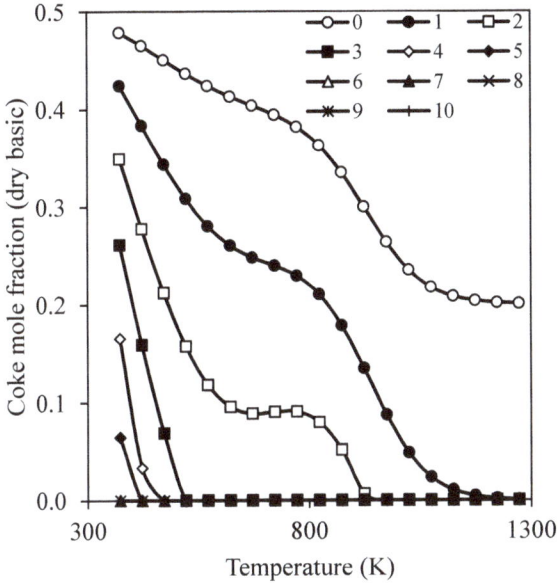

Figure 5. Coke formation at various steam to ethanol ratios and operating temperatures.

3.1.2. Sequential Steam Reformer and Shift Reactor (SRSR)

In this conventional reactor system (Figure 1a), the ethanol steam reformer and water gas shift reactor were carried out under the same operating condition. This can be considered as a single reactor for simulation in the Aspen Plus program.

In Figure 6a, the influence of reforming temperature and steam to ethanol ratio on the process efficiency is displayed. The reaction equilibrium shifted forward, resulting in higher hydrogen production and efficiency when increasing the steam to ethanol ratio. The process efficiency also improved with increasing reforming temperature up to 973 K. Beyond this temperature, its value reduced because the reverse water gas shift was the dominant reaction. Consequently, the best operating condition for SRSR was 973 K and a steam to ethanol ratio of 5, obtaining the highest efficiency at 67.50%. However, a large amount of carbon monoxide, ca. 14%, still remained in the product stream because the water gas shift reaction performed poorly at a high temperature. Thus, a water gas shift reaction that operated separately at lower temperature was suggested to reduce some carbon monoxide and improve the process efficiency. The reduction in reformed gas temperature from the ethanol steam reformer was proposed by quenching with water before flowing to the shift reactor. This will be discussed in the next section.

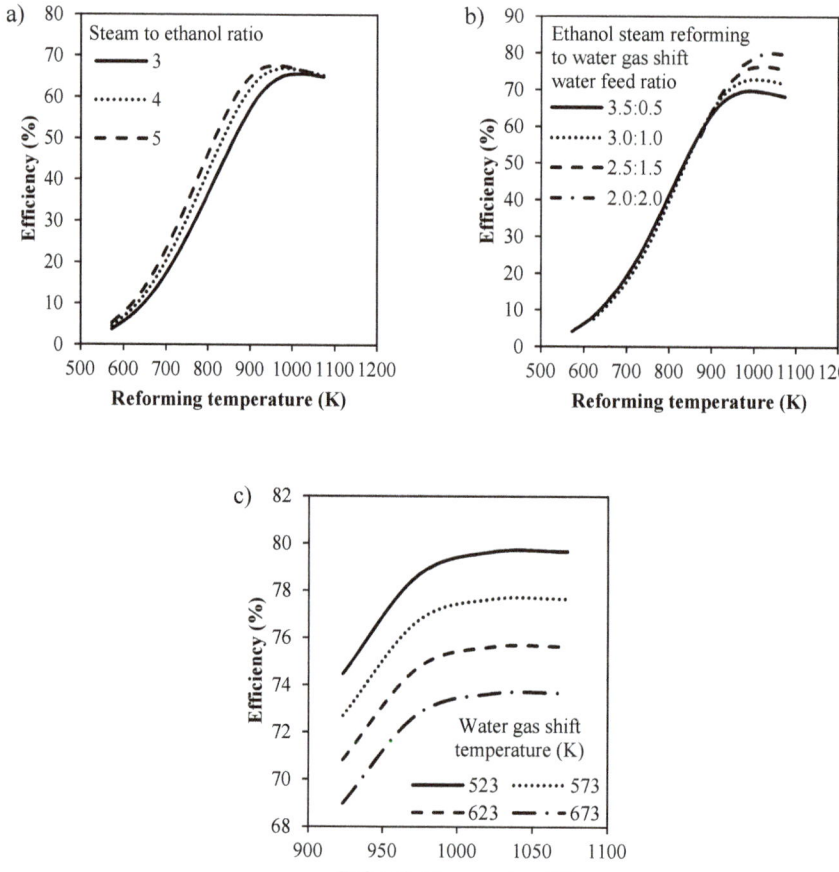

Figure 6. Process efficiency in percent of (**a**) the SRSR operated at various steam to ethanol ratios and reforming temperatures under a pressure of 1 atm, (**b**) SRSR-WS operated at various reforming temperatures and spiting ratios of stream under a pressure of 1 atm and a steam to ethanol ratio of 4, and (**c**) CHIRS operated at various reforming and water gas shift temperatures under a pressure of 1 atm and a steam to ethanol ratio of 4.

3.1.3. SRSR with Energy Management by Water Splitting (SRSR-WS)

To enhance the process efficiency, water was fed directly to the water gas shift reactor as shown in Figure 1b. According to the study of previous sections, the difference in processing efficiency at a steam to ethanol ratio between 4 and 5 was insignificant. Thus, the steam to ethanol ratio of 4 was selected due to less energy requirement. In the case of SRSR-WS, the overall steam to ethanol ratio of 4 was set as a constant value for the process and water was split for ethanol steam reforming and water gas shift at different ratios. For this study, the water feeding ratio at the ethanol steam reformer to water gas shift reactor varied between 3.5:0.5, 3.0:1.0, 2.5:1.5 and 2.0:2.0.

The efficiency of this system is shown as a function of temperature in Figure 6b. The system was considered without any presence of coke formation and condensation. Therefore, the water feeding ratio at the reformer lower than 2.0 was excluded due to the limitation of coking and condensation. Significant results obviously appeared when the reforming temperature was higher than 923 K. An improvement of process efficiency occurred when water was added to the water gas shift reactor at a higher ratio. The optimum condition was performed at the reforming temperature of 1023 K and the steam to ethanol ratio at the reformer to shift reactor of 2.0:2.0, achieving the highest efficiency of 79.90%. When compared to the SRSR process efficiency, its value for SRSR-WS improved significantly by about 12.40%. For the consideration of practical design of SRSR-WS, a direct feeding of water to the shift reactor may deactivate the catalyst and cause mechanical problems. Moreover, it was difficult to design and operate the system while injecting liquid into the reformed gas stream at the high temperature zone. Therefore, the SRSR-WS was not recommended. However, another concept was then created to solve this problem by considering a better energy management.

3.1.4. Compact Heat Integrated Reactor System (CHIRS)

A heat exchanger network was utilized to enhance the process efficiency within the ethanol steam reforming process. As exhibited in Figure 1c, two heat exchangers were integrated in the process to reduce the temperature of reformed gas and exhaust gas to be compatible with the water gas shift reactor and emission to environment, respectively. They were also used to preheat the reactant, ethanol and water, for hydrogen production. As investigated earlier, we preferred that the ethanol steam reformer was operated at a high operating temperature (923–1073 K), atmospheric pressure (1 atm), and steam to ethanol ratio of 4. Natural gas, assumed to contain methane (75%) and carbon dioxide (25%) was employed as a fuel for the combustor. The combustor supplied energy to the reformer for maintaining the reforming temperature. The exhaust gas leaving the process was limited to an outlet temperature of 523 K to avoid condensation. The reformed gas from the reformer was quenched to a low temperature between 523 and 673 K and then fed to the water gas shift reactor.

The results of process efficiency at various temperatures of reformer and shift reactor are reported in Figure 6c. The efficiency decreased with an increase in water gas shift temperature, while increasing the reforming temperature led to the enhancement of efficiency. When the reforming temperature was beyond 1023 K, the efficiency was almost constant. Hence, the reforming temperature of 1023 K was recommended. The highest efficiency of 79.65% was achieved when the process was operated at 1023 and 523 K for the ethanol steam reformer and water gas shift reactor, respectively. It was found that the efficiency of CHIRS was similar to that of SRSR-WS. However, CHIRS was the process with good energy management and can be operated practically. Therefore, this system was a proper design as a combined reactor consisting of a combustor, reformer and water gas shift reactor for further study in the next section.

3.2. Study of the Compact Heat Integrated Reactor System (CHIRS) via COMSOL Multiphysics

3.2.1. Preliminary Study of CHIRS

For the design of both CHIRS (I) and CHIRS (II), as displayed in Figures 2 and 3, the configuration dimension is summarized in Table 1. To simplify the simulation, the "stream in" to the combustor performed as a hot combusted gas with an inlet temperature of 1473 K. The ethanol flow rate was specified at 3.5 mmol/s with a steam to ethanol ratio of 4.

The influence of air and combusted gas on the reactor performance was investigated for the preliminary design of CHIRS. To find a proper air and combusted gas flow rate, CHIRS (II) was employed as a base case. The flow rate of air and combusted gas was varied between 106 and 1060 L/min.

In Figure 7a, at a higher air preheat flow rate in the air preheat chamber, a significant decrease in shift reactor temperature occurred while the reforming temperature was slightly reduced. The hydrogen fraction at the outlet stream of the process had the optimum value when air flow rate was 318 L/min. Since the steam reforming was a dominant reaction in the reformer, a lower reforming temperature led to less hydrogen production, even though the low shift reactor temperature could convert carbon monoxide and then boost hydrogen production. For the study of air flow rate, hydrogen could be produced in the range of 61–63% by changing the air flow rate from 106 to 1060 L/min.

Table 1. CHIRS (I) and CHIRS (II) dimensions.

Parameter	CHIRS (I)	CHIRS (II)	Unit
Combustion chamber inner diameter	234.644	234.644	mm
Combustion chamber outer diameter	240.644	240.644	mm
Air preheat chamber inner diameter	475.288	234.644	mm
Air preheat chamber outer diameter	481.288	240.644	mm
Reactor inner diameter	47.5	47.5	mm
Reactor outer diameter	53.5	53.5	mm
Ethanol steam reforming reactor length	300	300	mm
Water gas shift reactor length	300	300	mm
Air gap height	-	50	mm

The influence of combusted gas flow rate on the operating temperature of each section and hydrogen gained from the process is shown in Figure 7b. Average temperatures of both reactors were definitely enhanced with increasing the combusted gas flow rate. The steam reforming reaction rate in the reformer was absolutely improved at higher temperature, leading to an increase in hydrogen production. However, the hydrogen fraction was slightly lower when the combusted gas flow rate was beyond 530 L/min. The average shift reactor temperature was higher at the higher flow rate of combusted gas and limited the equilibrium of the water gas shift reaction after a complete conversion of ethanol and acetaldehyde. As a consequence, the combusted gas flow rate should be operated properly at 530 L/min to obtain the highest hydrogen production.

Air and combusted gas acted as a heat sink and heat source of the CHIRS. The reformer temperature decreased with increasing air flow rate, whereas it was increased by increasing the flow rate of the combusted gas. A better control of reformer and shift reactor temperatures to achieve hydrogen improvement was ascribed to the optimal air and combusted gas flow rate of 318 and 530 L/min, respectively. This condition was further used in computing the reactor performance of CHIRS (I) and CHIRS (II) as presented in Sections 3.2.2 and 3.2.3.

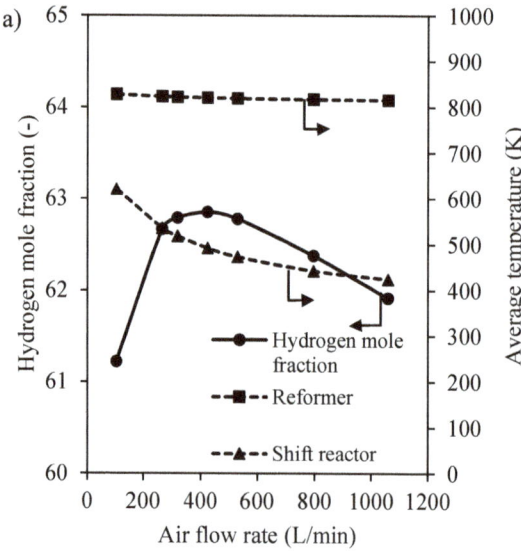

Figure 7. *Cont.*

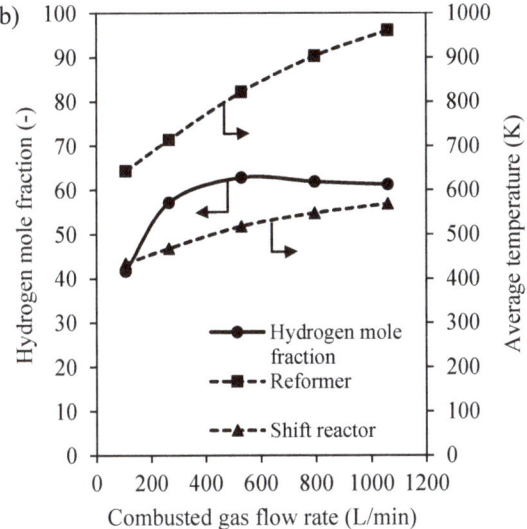

Figure 7. Influence of the (**a**) air preheat and (**b**) combusted gas flow rate on mole fraction of hydrogen and average temperature of the reformer and shift reactor.

3.2.2. Characteristics of CHIRS (I)

In the calculation domain inside the CHIRS (I) configuration, the temperature profile presentation as an isothermal contour is depicted in Figure 8. Combusted gas was intended to supply heat to the steam reformer, but heat was also conducted to the shift reactor through the wall between sections resulting in a higher temperature of the shift reactor. A small temperature gap between both sections occurred when the reactors were at 700–800 K for the reformer and 400–700 K for the water gas shift reactor (Figure 8b). For this CHIRS (I) design, the operating temperature control of each zone was poor, due to the presence of a low reforming temperature (752.15 K in average) and high shift reactor temperature (564.12 K in average).

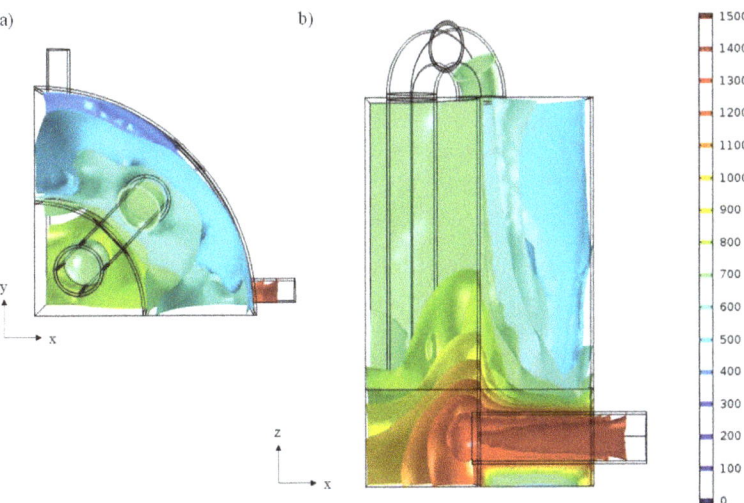

Figure 8. Temperature profile of CHIRS (I) in the (**a**) XY plane and (**b**) XZ plane.

Product distribution profiles indicated in terms of mole fraction for the reformer zone and shift reactor zone along with the reactor length are shown in Figure 9. According to the length of each zone as indicated in Table 1, the reformer part was considered in the range of 0.00–0.30 m followed by the water gas shift reactor in the range of 0.30–0.60 m. For the reformer, the related reactions, Equation (R1)–(R4), produce methane, acetaldehyde, carbon dioxide and syngas. In addition, ethanol was completely consumed within 0.1 m of length. Although the acetaldehyde steam reforming reaction (Equation (R4)) could occur inside the reformer, acetaldehyde still remained in the product stream due to low reaction rate at the low reforming temperature. After 0.3 m of length, the reformed gas entered the water gas shift section. The temperature was then reduced and the water gas shift reaction shifted forward, resulting in high hydrogen production of about 61.81% by mole with the presence of impurities such as methane, acetaldehyde, carbon monoxide and carbon dioxide.

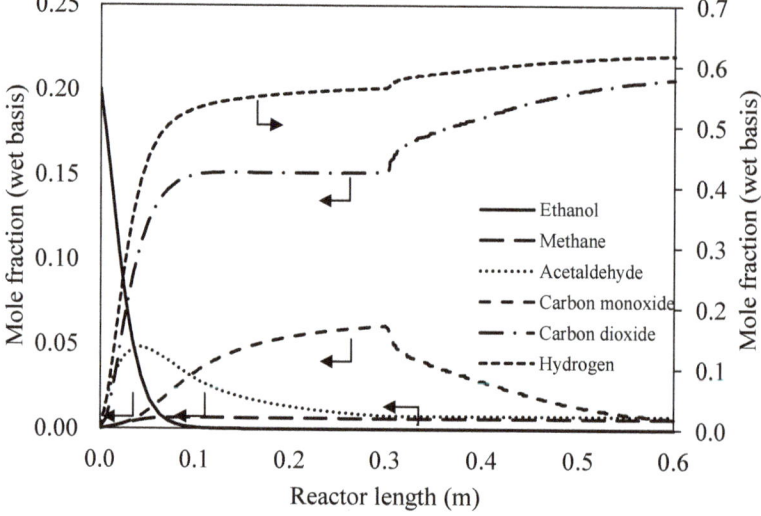

Figure 9. Mole fraction profiles at the center along CHIRS (I).

3.2.3. Characteristics of CHIRS (II)

In Figure 10, the temperature contour inside the CHIRS (II) configuration in YZ and XZ planes is presented. In this case, the combusted gas flowed through the inlet at the bottom of the reactor system. The uniform temperature profile appeared in the reforming section indicating a temperature between 700 and 900 K (Figure 10a). In Figure 10b, the green area at the reformer inlet representing a temperature around 700 K was apparent. It was likely due to highly consumed energy from a high reaction rate at the inlet. In the air gap section located between the reformer and the shift reactor, the reformed gas temperature was reduced to 600 K before entering the water gas shift section. Thus, the water gas shift reaction could perform at a temperature lower than 600 K. The air gap could greatly manage the temperature gradient between the combustion and air preheating chambers. As a result, the operating temperature was controlled perfectly for both the reforming reaction at high temperature and water gas shift reaction at low temperature.

According to this configuration (Table 1), mole fraction profiles of the product stream within the reformer zone (0.00–0.30 m) followed by the air gap section (0.30–0.35 m) and the water gas shift zone (0.35–0.65 m) distributing along the reactor length are shown in Figure 11. Owing to the high reforming temperature, ethanol and acetaldehyde were totally consumed within the ethanol steam reformer section. A reverse water gas shift also occurred at the middle of the reforming section as the mole fraction of carbon dioxide was reduced. The reformed gas then flowed through the air gap section without any further reactions. Hence, the same gas composition entered the water gas shift section. Inside the shift reactor zone, the water gas shift reaction evidently took place. Carbon monoxide sharply decreased when hydrogen was noticeably increased up to 63.48% (wet basis). The product stream mainly comprised hydrogen and carbon dioxide with a very low fraction of methane and carbon monoxide.

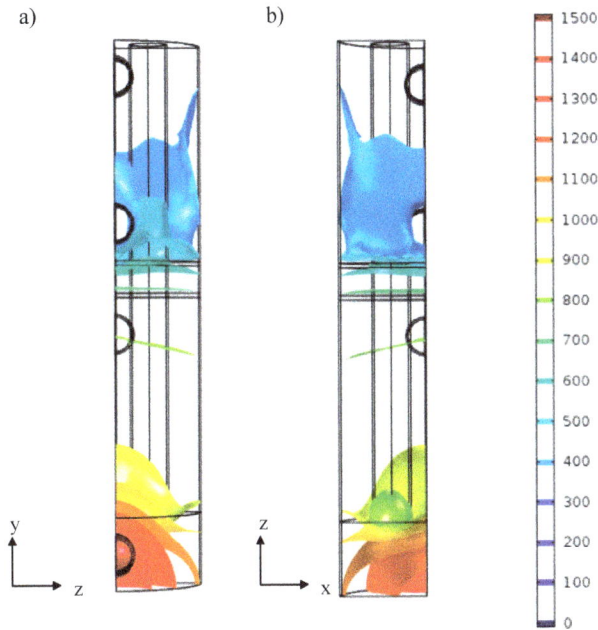

Figure 10. Temperature profile of CHIRS (II) in the (**a**) YZ plane and (**b**) XZ direction.

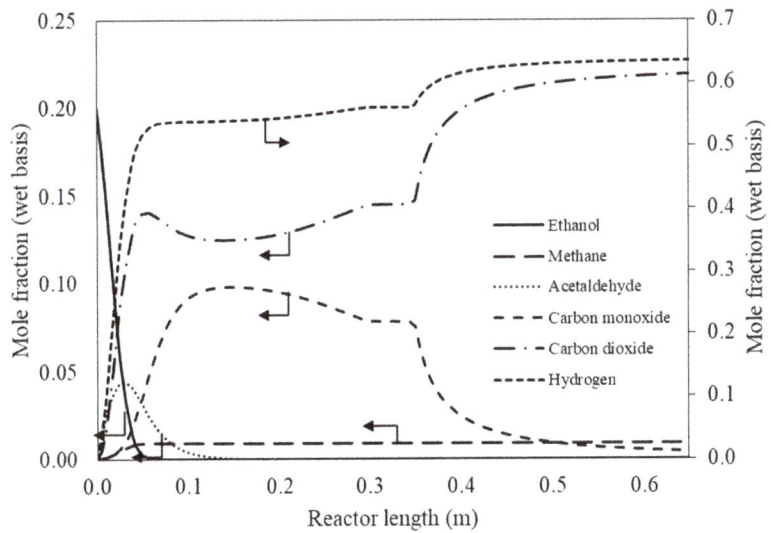

Figure 11. Mole fraction profiles at the center along CHIRS (II).

3.2.4. Comparison in Process Performance between CHIRS (I) and CHIRS (II)

A compact heat integrated reactor system (CHIRS) combining combustor, reformer and shift reactor was designed to produce hydrogen for a stationary system with two different configurations. An overview of CHIRS (I) and CHIRS (II), including reactor volume and other reactor performance, is listed and summarized in Table 2. It was obviously found that the required volume space for CHIRS (I) was considerably higher than that for CHIRS (II), indicating the value of 72.77 and 34.11 L, respectively. The temperature gradient between reformer and shift reactor was smaller in CHIRS (I), when compared to another design. These average temperature differences between two reactors were 188.03 and 323.50 K, respectively. At a higher temperature gradient between the reformer and shift reactor, a more appropriate operating temperature with higher reformer temperature and lower shift reactor temperature led to a better yield of hydrogen. Therefore, CHIRS (II) provided higher hydrogen selectivity with an absence of acetaldehyde in the product stream. In addition, hydrogen was produced at 15.36 mmol/s, which was equivalent to 4.39 kW of hydrogen energy. The CHIRS (I) design obtained more hydrogen production, which was about 16.47 mmol/s (equivalent to 4.71 kW of hydrogen energy). However, the hydrogen production was insignificantly different. From another point of view, the volume of CHIRS (II) was half of that for CHIRS (I), whereas CHIRS (II) achieved a higher hydrogen selectivity than the other one. Moreover, the CHIRS (II) design could efficiently manage the temperature of each reactor in the combined system. Eventually, this design was recommended for hydrogen production from ethanol. However, it should be noted that when applying the hydrogen production system for PEM application, the concentration of CO in the hydrogen rich gas needs to be considered and the operating condition needs to be carefully selected to comply with the CO tolerance in PEM.

Table 2. Process performance of CHIRS (I) and CHIRS (II).

Reactor Performance	CHIRS (I)	CHIRS (II)	Unit
Reactor volume	72.77	34.11	L
Average reforming temperature	752.15	805.44	K
Average water gas shift temperature	564.12	481.94	K
Carbon monoxide selectivity	0.61	0.43	%
Methane selectivity	0.62	0.91	%
Acetaldehyde selectivity	0.77	0.00	%
Hydrogen selectivity	61.81	63.49	%
Hydrogen production	16.47	15.36	mmol/s

4. Conclusions

The compact heat integrated reactor system (CHIRS) consisting of a combustor, a reformer and a water gas shift reactor was finally developed for hydrogen production from ethanol. Firstly, three reformer concepts, including SRSR, SRSR-WS, and CHIRS, relying on the level of concept development were investigated via the Aspen Plus program. According to the high process efficiency, the condition was suitably operated at a reforming temperature of 573–1073 K, reforming pressure of 1 atm and steam to ethanol ratio in the range of 3–5. The SRSR was considered as a conventional process producing a high composition of carbon monoxide with high heat loss at the outlet, resulting in low efficiency (67.50%). The SRSR-WS and CHIRS could improve the process efficiency up to 79.90% and 79.65%, respectively, due to heat integration within the system. Since the SRSR-WS had some trouble with regard to practical design and operation, the promising CHIRS design was selected to be studied in detail via COMSOL Multiphysics software. For the CHIRS design, the reformer and shift reactor were placed inside the combustion and air preheating chambers, respectively. Both chambers were integrated in the radial direction for the CHIRS (I) design and the vertical direction for the CHIRS (II) design. In the preliminary step, the CHIRS design was examined by varying the air and combusted gas flow rate to obtain a high reactor performance. It was found that the air and combusted gas flow rate was 318 and 530 L/min, respectively. The CHIRS (I) design provided a lower reformer temperature and higher shift reactor temperature, compared to the CHIRS (II) design. A temperature control in the CHIRS (II) design was better, resulting in a higher reactor performance in each section. Additionally, the CHIRS (II) design occupied half the volume of the CHIRS (I) design, whereas it provided similar hydrogen production with higher hydrogen selectivity. The CHIRS (II) design was ultimately recommended as a stationary combined reactor for hydrogen production from ethanol.

Author Contributions: Conceptualization, S.C., S.A., W.K., J.C. and T.J.; simulation, preliminary data analysis and writing—original draft preparation, T.J. and W.K.; writing—review and editing, W.K., K.O., S.W., S.C. and S.A.; funding acquisition, S.C. and S.A. All authors have read and agreed to the published version of the manuscript.

Funding: The supports from the Royal Golden Jubilee. Program from the Thailand Research Fund and Chulalongkorn University, National Metal and Materials Technology Center (MTEC), NSTDA and the "Research Chair Grant" National Science and Technology Development Agency (NSTDA) are gratefully acknowledged.

Acknowledgments: The authors would like to thank Navadol Laosiripojana from The Joint Graduate School of Energy and Environment and Pisanu Toochinda from Sirindhorn International Institute of Technology for their useful suggestions.

Conflicts of Interest: The authors declare no conflict of interest.

Nomenclature

C_p	Specific heat at constant pressure (J kg^{-1} K^{-1})
$D_{i,eff}$	Mass diffusion coefficient of species i in mixture (m^2 s^{-1})
g	Gravity force (m/s^2)
ΔH_j	Heat of reaction j (J/mol)
$\Delta H_{c,i}$	Heat of combustion of species i (kW mol^{-1})
k	Thermal conductivity (W m^{-1} K^{-1})
K_{WGS}	Thermodynamics equilibrium constant (-)
MW_i	Molar mass of species i (kg kmol^{-1})
$\dot{n}$	Mole flow rate (mol s^{-1})
p	Pressure (Pa)
r_j	Rate of reaction j (mol s^{-1} m^{-2})
T	Temperature (K)
$\vec{v}$	Velocity (m s^{-1})

Greek symbols

ρ	Density (kg m^{-3})
μ	Dynamic viscosity (Pa s)
ω_i	Mass fraction of species i

References

1. Liu, K.; Song, C.; Subramani, V. *Hydrogen and Syngas Production and Purification Technologies*; Wiley Online Library: Hoboken, NJ, USA, 2010.
2. Padró, C.; Lau, F. *Advances in Hydrogen Energy*; Springer: Berlin/Heidelberg, Germany, 2000.
3. Dincer, I.; Acar, C. Review and evaluation of hydrogen production methods for better sustainability. *Int. J. Hydrog. Energy* **2015**, *40*, 11094–11111. [CrossRef]
4. Mehrpooya, M.; Moftakhari Sharifzadeh, M.M.; Rajabi, M.; Aghbashlo, M.; Tabatabai, M.; Hosseinpour, S.; Ramakrishna, S. Design of an integrated process for simultaneous chemical looping hydrogen production and electricity generation with CO$_2$ capture. *Int. J. Hydrog. Energy* **2017**, *42*, 8486–8496. [CrossRef]
5. Nikolaidis, P.; Poullikkas, A. A comparative overview of hydrogen production processes. *Renew. Sustain. Energy Rev.* **2017**, *67*, 597–611. [CrossRef]
6. Nahar, G.; Mote, D.; Dupont, V. Hydrogen production from reforming of biogas: Review of technological advances and an Indian perspective. *Renew. Sustain. Energy Rev.* **2017**, *76*, 1032–1052. [CrossRef]
7. Xue, Y.-P.; Yan, C.-F.; Zhao, X.-Y.; Huang, S.-L.; Guo, C.-Q. Ni/La$_2$O$_3$-ZrO$_2$ catalyst for hydrogen production from steam reforming of acetic acid as a model compound of bio-oil. *Korean J. Chem. Eng.* **2017**, *34*, 305–313. [CrossRef]

8. Quan, C.; Xu, S.; Zhou, C. Steam reforming of bio-oil from coconut shell pyrolysis over Fe/olivine catalyst. *Energy Convers. Manag.* **2017**, *141*, 40–47. [CrossRef]
9. Nabgan, W.; Tuan Abdullah, T.A.; Mat, R.; Nabgan, B.; Gambo, Y.; Ibrahim, M.; Ahmad, A.; Jalil, A.A.; Triwahyono, S.; Saeh, I. Renewable hydrogen production from bio-oil derivative via catalytic steam reforming: An overview. *Renew. Sustain. Energy Rev.* **2017**, *79*, 347–357. [CrossRef]
10. Haryanto, A.; Fernando, S.; Murali, N.; Adhikari, S. Current Status of Hydrogen Production Techniques by Steam Reforming of Ethanol: A Review. *Energy Fuels* **2005**, *19*, 2098–2106. [CrossRef]
11. Ni, M.; Leung, D.Y.C.; Leung, M.K.H. A review on reforming bio-ethanol for hydrogen production. *Int. J. Hydrog. Energy* **2007**, *32*, 3238–3247. [CrossRef]
12. Vaidya, P.D.; Rodrigues, A.E. Insight into steam reforming of ethanol to produce hydrogen for fuel cells. *Chem. Eng. J.* **2006**, *117*, 39–49. [CrossRef]
13. Hou, T.; Zhang, S.; Chen, Y.; Wang, D.; Cai, W. Hydrogen production from ethanol reforming: Catalysts and reaction mechanism. *Renew. Sustain. Energy Rev.* **2015**, *44*, 132–148. [CrossRef]
14. Sharma, Y.C.; Kumar, A.; Prasad, R.; Upadhyay, S.N. Ethanol steam reforming for hydrogen production: Latest and effective catalyst modification strategies to minimize carbonaceous deactivation. *Renew. Sustain. Energy Rev.* **2017**, *74*, 89–103. [CrossRef]
15. Dou, B.; Zhang, H.; Cui, G.; Wang, Z.; Jiang, B.; Wang, K.; Chen, H.; Xu, Y. Hydrogen production and reduction of Ni-based oxygen carriers during chemical looping steam reforming of ethanol in a fixed-bed reactor. *Int. J. Hydrog. Energy* **2017**, *42*, 26217–26230. [CrossRef]
16. Tripodi, A.; Compagnoni, M.; Ramis, G.; Rossetti, I. Process simulation of hydrogen production by steam reforming of diluted bioethanol solutions: Effect of operating parameters on electrical and thermal cogeneration by using fuel cells. *Int. J. Hydrog. Energy* **2017**, *42*, 23776–23783. [CrossRef]
17. Castedo, A.; Uriz, I.; Soler, L.; Gandía, L.M.; Llorca, J. Kinetic analysis and CFD simulations of the photocatalytic production of hydrogen in silicone microreactors from water-ethanol mixtures. *Appl. Catal. B Environ.* **2017**, *203*, 210–217. [CrossRef]
18. Wang, W.; Wang, Y.Q. Thermodynamic analysis of steam reforming of ethanol for hydrogen generation. *Int. J. Energy Res.* **2008**, *32*, 1432–1443. [CrossRef]
19. Rabenstein, G.; Hacker, V. Hydrogen for fuel cells from ethanol by steam-reforming, partial-oxidation and combined auto-thermal reforming: A thermodynamic analysis. *J. Power Sources* **2008**, *185*, 1293–1304. [CrossRef]
20. Wanat, E.C.; Venkataraman, K.; Schmidt, L.D. Steam reforming and water–gas shift of ethanol on Rh and Rh–Ce catalysts in a catalytic wall reactor. *Appl. Catal. A Gen.* **2004**, *276*, 155–162. [CrossRef]
21. Utaka, T.; Okanishi, T.; Takeguchi, T.; Kikuchi, R.; Eguchi, K. Water gas shift reaction of reformed fuel over supported Ru catalysts. *Appl. Catal. A Gen.* **2003**, *245*, 343–351. [CrossRef]
22. Amadeo, N.E.; Laborde, M.A. Hydrogen production from the low-temperature water-gas shift reaction: Kinetics and simulation of the industrial reactor. *Int. J. Hydrog. Energy* **1995**, *20*, 949–956. [CrossRef]
23. Pala, L.P.R.; Wang, Q.; Kolb, G.; Hessel, V. Steam gasification of biomass with subsequent syngas adjustment using shift reaction for syngas production: An Aspen Plus model. *Renew. Energy* **2017**, *101*, 484–492. [CrossRef]
24. Kaftan, A.; Kusche, M.; Laurin, M.; Wasserscheid, P.; Libuda, J. KOH-promoted Pt/Al$_2$O$_3$ catalysts for water gas shift and methanol steam reforming: An operando DRIFTS-MS study. *Appl. Catal. B Environ.* **2017**, *201*, 169–181. [CrossRef]
25. Andisheh Tadbir, M.; Akbari, M.H. Integrated methanol reforming and oxidation in wash-coated microreactors: A three-dimensional simulation. *Int. J. Hydrog. Energy* **2012**, *37*, 2287–2297. [CrossRef]
26. Grote, M.; Maximini, M.; Yang, Z.; Engelhardt, P.; Köhne, H.; Lucka, K.; Brenner, M. Experimental and computational investigations of a compact steam reformer for fuel oil and diesel fuel. *J. Power Sources* **2011**, *196*, 9027–9035. [CrossRef]
27. Yin, F.; Ji, S.; Mei, H.; Zhou, Z.; Li, C. Coupling of highly exothermic and endothermic reactions in a metallic monolith catalyst reactor: A preliminary experimental study. *Chem. Eng. J.* **2009**, *155*, 285–291. [CrossRef]
28. Jiwanuruk, T.; Putivisutisak, S.; Ponpesh, P.; Kositanont, C.; Tagawa, T.; Yamada, H.; Fukuhara, C.; Assabumrungrat, S. Comparison between parallel and checked arrangements of micro reformer for H$_2$ production from methane. *Chem. Eng. J.* **2015**, *268*, 135–143. [CrossRef]

29. Seo, Y.-S.; Seo, D.-J.; Seo, Y.-T.; Yoon, W.-L. Investigation of the characteristics of a compact steam reformer integrated with a water-gas shift reactor. *J. Power Sources* **2006**, *161*, 1208–1216. [CrossRef]
30. Hayer, F.; Bakhtiary-Davijany, H.; Myrstad, R.; Holmen, A.; Pfeifer, P.; Venvik, H.J. Synthesis of dimethyl ether from syngas in a microchannel reactor—Simulation and experimental study. *Chem. Eng. J.* **2011**, *167*, 610–615. [CrossRef]
31. Lima da Silva, A.; Malfatti, C.d.F.; Müller, I.L. Thermodynamic analysis of ethanol steam reforming using Gibbs energy minimization method: A detailed study of the conditions of carbon deposition. *Int. J. Hydrogen Energy* **2009**, *34*, 4321–4330. [CrossRef]
32. Uriz, I.; Arzamendi, G.; López, E.; Llorca, J.; Gandía, L.M. Computational fluid dynamics simulation of ethanol steam reforming in catalytic wall microchannels. *Chem. Eng. J.* **2011**, *167*, 603–609. [CrossRef]
33. Montero, C.; Remiro, A.; Valle, B.; Oar-Arteta, L.; Bilbao, J.; Gayubo, A.G. Origin and Nature of Coke in Ethanol Steam Reforming and Its Role in Deactivation of Ni/La2O3–αAl2O3 Catalyst. *Ind. Eng. Chem. Res.* **2019**, *58*, 14736–14751. [CrossRef]

© 2020 by the authors. Licensee MDPI, Basel, Switzerland. This article is an open access article distributed under the terms and conditions of the Creative Commons Attribution (CC BY) license (http://creativecommons.org/licenses/by/4.0/).

Article

Effect of CuO as Sintering Additive in Scandium Cerium and Gadolinium-Doped Zirconia-Based Solid Oxide Electrolysis Cell for Steam Electrolysis

R. Visvanichkul [1], S. Peng-Ont [2], W. Ngampuengpis [2], N. Sirimungkalakul [2], P. Puengjinda [2], T. Jiwanuruk [2], T. Sornchamni [2] and P. Kim-Lohsoontorn [1,*]

[1] Centre of Excellence on Catalysis and Catalytic Reaction Engineering, Department of Chemical Engineering, Faculty of Engineering, Chulalongkorn University, Bangkok 10330, Thailand; r.visvanichkul@gmail.com
[2] PTT Innovation Institute, PTT Public Company Limited, Ayutthaya 13170, Thailand; saranya.p@pttplc.com (S.P.-O.); Ng.watcharin@gmail.com (W.N.); Nkanjanabat@gmail.com (N.S.); pramote.pu@pttplc.com (P.P.); tara.j@pttplc.com (T.J.); thana.s@pttplc.com (T.S.)
* Correspondence: pattaraporn.k@chula.ac.th

Received: 31 August 2019; Accepted: 18 November 2019; Published: 21 November 2019

Abstract: The effect of CuO as a sintering additive on the electrolyte of solid oxide electrolysis cells (SOECs) was investigated. 0.5 wt% CuO was added into $Sc_{0.1}Ce_{0.05}Gd_{0.05}Zr_{0.89}O_2$ (SCGZ) electrolyte as a sintering additive. An electrolyte-supported cell (Pt/SCGZ/Pt) was fabricated. Phase formation, relative density, and electrical conductivity were investigated. The cells were sintered at 1373 K to 1673 K for 4 h. The CuO significantly affected the sinterability of SCGZ. The SCGZ with 0.5 wt% CuO achieved 95% relative density at 1573 K while the SCGZ without CuO could not be densified even at 1673 K. Phase transformation and impurity after CuO addition were not detected from XRD patterns. Electrochemical performance was evaluated at the operating temperature from 873 K to 1173 K under steam to hydrogen ratio at 70:30. Adding 0.5 wt% CuO insignificantly affected the electrochemical performance of the cell. Activation energy of conduction (E_a) was 72.34 kJ mol^{-1} and 74.93 kJ mol^{-1} for SCGZ and SCGZ with CuO, respectively.

Keywords: solid oxide electrolysis cells; sintering additive; CuO; hydrogen production; steam electrolysis

---

1. Introduction

Global climate change is an important issue that is clearly found to affect the environment and humanity. Greenhouse gas emission is the major cause for retaining heat in the atmosphere, leading to climate change. Greenhouse gases are mainly generated from the combustion of fossil fuel such as petroleum, natural gas, and coal. Hydrogen is the one of the promising energy carriers for better environment since only water and energy are produced from hydrogen combustion [1,2]. Moreover, hydrogen can be used as a feedstock for various industrial chemical productions [3–5].

Solid oxide electrolysis cells (SOECs) are used to produce high purity hydrogen from steam electrolysis reaction at elevated temperatures. Material selection is one of the challenges faced in the implementation of SOEC. The operation may operate under severe condition in the production scale. Sealant of gas leakage between the stack needs to be modified [6]. Material choice for SOEC electrode is often similar to solid oxide fuel cell [7–9]. Therefore, 8 mol% yttria stabilized zirconia (YSZ), which is a conventional electrolyte in solid oxide fuel cell [10–13], is generally used as the SOEC electrolyte. Attempt has been made to seek for fast ionic conductor instead of conventional YSZ [14–18]. Doped-cerate- or doped-zirconia-based electrolyte has been studied. Scandium oxide (Sc_2O_3) is the one among the dopants for zirconia-based electrolyte (Scandium stabilized zirconia, ScSZ) that improves the ionic conductivity when compared with YSZ electrolyte. Similar ionic radius between Sc^{3+} and Zr^{4+} can cause low internal stress and the activation energy of conduction [16]. However, phase transition can occur in ScSZ electrolyte, leading to loss of conductivity. Abbas et al. [19] prepared co-doping of 1 mol% Gd_2O_3, CaO, and CeO_2 with 10 mol% Sc_2O_3 into zirconia structure to reduce phase transition and it was found that 1 mol% CeO_2 and 10 mol% Sc_2O_3 as dopants provide relatively highest conductivity, likely because of the close ionic radius of Ce^{4+} and Zr^{4+} compared to other dopants. Shin et al. [20] investigated the co-doping of trivalent oxides (Gd_2O_3, Yb_2O_3, and Y_2O_3) with CeO_2 into scandium stabilized zirconia structure. It was reported that co-doping of Gd_2O_3 with CeO_2 provides relatively higher ionic conductivity when compared with other oxides. However, doped-zirconia electrolyte often requires relatively high sintering temperature (~1723 K) for electrolyte densification. High sintering temperature of the electrolyte can lead to degradation issue in other cell compartments, leading to difficulty in fabrication process. For example, in electrode-supported cells, nickel (Ni) component in an electrode can agglomerate at high temperature, leading to decreasing three phase boundary of the electrode. In metal-supported cell, metal part (often containing iron (Fe) or Ni content) cannot withstand high sintering temperature because of a large thermal expansion coefficient of the metal. Therefore, lower sintering temperature for electrolyte is required to enhance the fabrication process of the cell.

Sintering additive can promote densification and help reduce sintering temperature of the electrolyte. The additive can melt to form liquid phase during the sintering process. Solid grain of the electrolyte is wet by liquid phase which provide capillary force pulling the grain denser [16]. There are varied metal oxide such as CuO, NiO, Fe_2O_3, and MnO_2 that have been used as sintering additives for zirconia-based electrolyte [21–24]. However, adding sintering additives which are often metal oxides can induce electronic conductivity in the electrolyte. Myung et al. [25] investigated scanning electron microscope (SEM) images of sintered YSZ by varying amount of CuO additive from 0 to 1.5 wt%. It was found that optimum quantity of additive and sintering temperature was 0.5 wt% of CuO and 1623 K, respectively. Our previous work [26] reported other metal oxides (NiO, Co_2O_3, and ZnO) as additives in barium cerate-based electrolyte. The additives helped increase sinterability of the electrolyte and provided different conductivity for the electrolyte.

In this research, $Sc_{0.1}Ce_{0.05}Gd_{0.05}Zr_{0.89}O_2$ (SCGZ) was used as an electrolyte in SOEC. CuO was added into the electrolyte as a sintering additive. The effect of sintering additive on phase formation, cell densification, and electrical performance were investigated.

2. Experimental

2.1. Cell Fabrication

Scandium cerium- and gadolinium-doped zirconia ($Sc_{0.1}Ce_{0.05}Gd_{0.05}Zr_{0.89}O_2$, SCGZ) was fabricated into an electrolyte-supported SOEC. Three grams of SCGZ powder (Kceracell, Chungcheongnam-do, Republic of Korea) were pressed at 24 MPa into a pellet with diameter of 25 mm and thickness of 1.4 mm. The pellets were then sintered at 1373 K to 1673 K for 4 h. The electrolyte pellet with sintering additive was fabricated using the same method when 0.5 wt% CuO was added into the SCGZ staring powder. The relative density of the sintered electrolyte pellet was calculated following Equation (1).

$$\text{Relative density} = \frac{\text{Actual density}}{\text{Theoretical density}} \times 100 \qquad (1)$$

where actual density is the bulk density of sintered cell. The theoretical density was obtained by lattice parameter from X-ray diffraction (XRD) analysis (Bruker D8 Advance, Billerica, MA, USA).

Platinum (Pt) conductive paste (70 wt% Pt, Nexceris) was coated on the both sides of the electrolyte pellet as electrodes with thickness and area of ~30 μm and ~0.5 cm^2, respectively. The schematic of cell component was shown as Figure 1. The cell was then fired at 1173 K for 2 h. It should be noted that the choices of electrode used depend on the deposition technique, the operating temperature, and the type of conductivity. Using gold or platinum as working electrode was reported to perform well at high temperature but below 873 K they are relatively blocking oxygen [27].

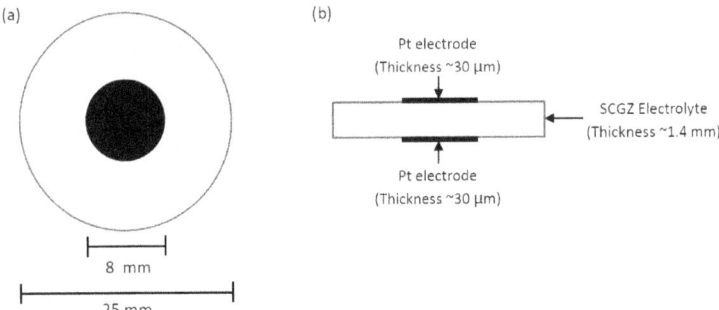

Figure 1. Schematic drawing of electrolyte-supported solid oxide electrolysis cells (SOEC); (**a**) top view, (**b**) side view.

2.2. Characterization

Phase and crystallite size of sintered electrolyte were determined using XRD with CuKα source (Bruker D8 Advance, Billerica, MA, USA). The microstructure of the samples was investigated using SEM (Hitachi S-3400N JEOL model S-3400, Tokyo, Japan). The average grain sizes were calculated by linear intersection method.

2.3. Electrochemical Performance Measurement

Electrochemical performance was measured in a controlled temperature from 873 K to 1173 K with the feed containing steam and hydrogen at the ratio of 70:30. The fabricated cell was attached with Pt mesh and wire (Kceracell, Chungcheongnam-do, Korea) for electrical connection. The cell was placed on a cell holder with high temperature sealant (Ceramabond 552, Aramco, Houston, TX, USA). The holder was installed inside a vertical furnace (Chavachote, Bangkok, Thailand). High performance liquid chromatography (HPLC) liquid pump (Teledyne SSI, State College, PA, USA) was used to supply deionized water through a heated-pipe for steam generation in the system. Linear sweep voltammetry procedure was applied to generate current/voltage (I/V) curves by controlling the potential from 0.4 V to 1.8 V with a scan rate of 20 mVs^{-1} (Metrohm Autolab, Utrecht, The Netherlands).

Resistance was determined by the slope of current/voltage (I/V) curves and the conductivity of fabricated cell (σ) was then calculated using Equation (2).

$$\sigma = L/RA \quad (2)$$

where σ is the conductivity (S cm^{-1}); L is the thickness of fabricated cell (cm); R is the cell resistance (Ω); and, A is the area of electrode (cm^2).

The activation energy of conduction (E_a) was obtained by using Arrhenius, Equation (3) with conductivity value as mentioned above.

$$\sigma T = A \cdot \exp\left(-\frac{E_a}{RT}\right) \quad (3)$$

where σ is the conductivity (S cm^{-1}); T is the absolute temperature (K); A is a constant; E_a is the activation energy of conduction (J mol^{-1}); and R is the gas constant (8.314 J K^{-1} mol^{-1}). It should be noted that the activation energy of conduction in this study was calculated from I/V curves ranging from open circuit voltage to 1.8 V. The slope of I/V curve is total resistance, which includes electrode resistance. The I/V slope was not constant and was derived using a linear regression with R-squared

(R^2) ranging from 0.90–0.97. In this study, Pt was applied as both electrodes (Pt/Electrolyte/Pt) and was expected to provide rather low resistance at operating conditions.

3. Results and Discussion

3.1. Densification of the Fabricated Cell

SEM images of SCGZ electrolyte with and without 0.5 wt% CuO sintered at varied temperature from 1423 K to 1673 K are presented in Figures 2 and 3, respectively. The microstructure images reveal that the SCGZ without the sintering additive could not be densified although high sintering temperature was used. Porosity were observed all over the SCGZ without the sintering additive. Grain boundary was observed from 1623 K but the grain size was rather small. The porosity decreased when the sintering temperature was increased. However, the relative density of the SCGZ without the sintering additive was only <90%, although high sintering temperature was increased up to 1673 K. On the other hand, densification and larger grain size were observed in the sample with the sintering additive. The grain growth was observed when increasing sintering temperature (Table 1). The added CuO could diffuse along the grain boundary and substituted in the vacancy position of the microstructure. Myung et al. [25] investigated CuO (0.3 wt% to 1.5 wt%) as sintering additive in yttria-stabilized zirconia (YSZ) sintering between 1423 K to 1673 K. It was reported that 0.5 wt% of CuO is the optimal amount providing the highest densification. Amount of optimal CuO additive at 0.5 wt% was the same for YSZ and SCGZ, likely because both materials are zirconia-based electrolyte. However, although having same optimal CuO amount, YSZ requires relatively higher sintering temperature (1623 K) when compared with SCGZ (1573 K) at the same sintering additive amount.

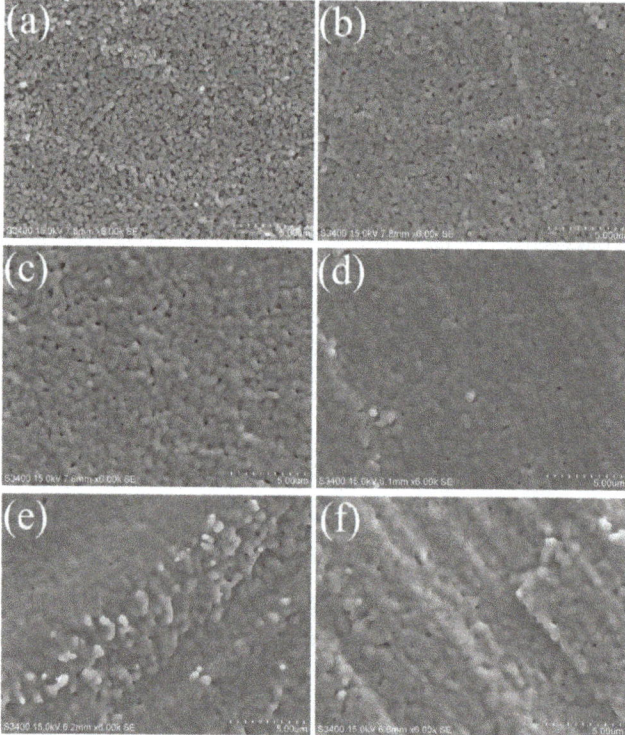

Figure 2. Scanning electron microscope (SEM) images of sintered $Sc_{0.1}Ce_{0.05}Gd_{0.05}Zr_{0.89}O_2$ (SCGZ) at (**a**) 1423 K, (**b**) 1473 K, (**c**) 1523 K, (**d**) 1573 K, (**e**) 1623 K, and (**f**) 1673 K.

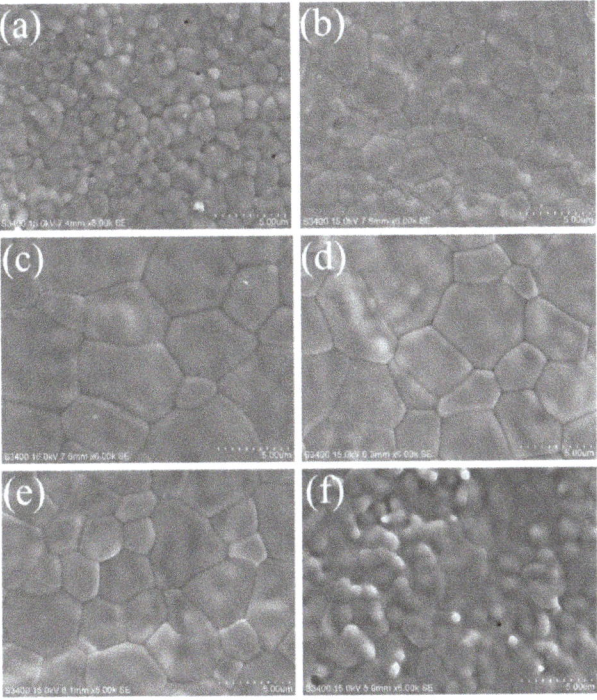

Figure 3. SEM images of sinterted SCGZ with 0.5 wt% CuO at (**a**) 1423 K, (**b**) 1473 K, (**c**) 1523 K, (**d**) 1573 K, (**e**) 1623 K and (**f**) 1673 K.

Table 1. Average grain size of sinterted $Sc_{0.1}Ce_{0.05}Gd_{0.05}Zr_{0.89}O_2$ (SCGZ) electrolyte and sintered SCGZ electrolyte with 0.5 wt% CuO when sintering temperature was varied.

Sintering Temperature (K)	Average Grain Size (µm)	
	SCGZ	SCGZ with 0.5 wt% CuO
1423	-	1.58
1473	-	2.66
1523	1.04	5.02
1573	1.20	5.11
1623	1.51	3.32
1673	2.28	3.77

Adding 0.5 wt% CuO enhanced the densification of SCGZ. The relative densities of the fabricated cells are presented in Figure 4. The SCGZ without sintering additive provided rather low relative density (<90%) at all sintering temperatures, corresponding to porosity observed in the SEM images. The SCGZ with 0.5 wt% CuO could be densified at lower sintering temperature. Increasing sintering temperature from 1423 to 1673 K could help increase the relative density of the fabricated cell. The cell was densified at 95% relative density at 1573 K respect to other sintering temperature. It was reported that CuO exhibit relatively low melting point (around 1599 K) and can enhance sintering by pore filling during liquid phase sintering [28]. Increasing temperature above 1623 K was found to decrease the relative density of the fabricated cell, likely relating to the liquid phase sintering. Liou et al. [29] studied the effect of CuO as sintering additives on $CaTiO_3$ perovskite ceramics. Liquid phase sintering at grain boundary is found at sintered sample at 1723 K for 6 h and increases significantly when

increasing sintering temperature to 1743 K. However, liquid phase did not occur when increasing sintering temperature up to 1773 K for 6 h, leading to less densification. Moreover, when sintering soak time was increased to 8 h at 1723 K, liquid phase sintering significantly increases when compared with 6 h. A proper sintering temperature and soak time are important factors affecting the densification of the sample.

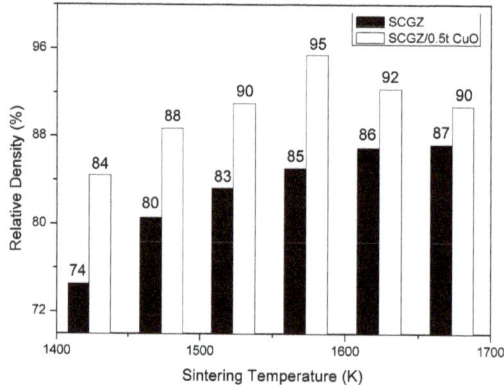

Figure 4. Relative density (%) of the SCGZ pellet and the SCGZ pellet with 0.5 wt% CuO when the sintering temperature was varied.

3.2. Phase Identification

XRD patterns of the SCGZ pellet sintered at 1673 K and the SCGZ with 0.5 wt% CuO pellet sintered at 1523 K are shown in Figure 5. The XRD patterns included main peaks at (111), (200), (220), (311), and (222) planes (COD Database ID: 1529100). Impurity phase was not detected. The XRD patterns of two samples were identical in term of the peak positions. Shifting in peaks position of the XRD patterns was not detected. This result could confirm that CuO did not form into a solid solution with SCGZ but well-mixed with SCGZ as a composite form. This results corresponded to the previous work [30]. It should be noted that CuO peaks were not detected, likely due to small amount of CuO in the sample. The average crystallite sizes were 184 nm and 202 nm for SCGZ sintered at 1673 K and SCGZ with sintering additive sintered at 1523 K, respectively, corresponding to the work that reported higher densification provided a larger grain size leading to a larger crystallite size [31].

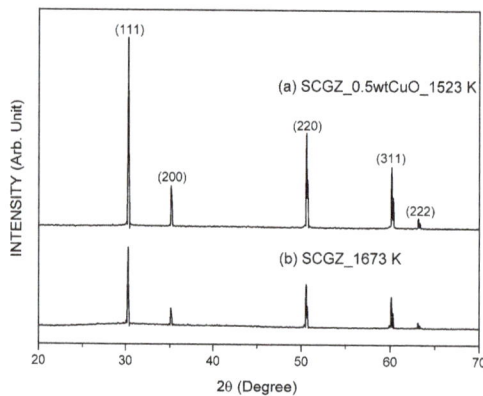

Figure 5. The X-ray diffraction patterns of (a) sintered SCGZ at 1673 K and (b) sintered SCGZ with 0.5 wt% CuO at 1523 K.

3.3. Activation Energy of Conduction

The electrochemical performance of the SOEC was evaluated. Linear sweep voltammetry was conducted from 873 K to 1173 K under a constant steam to hydrogen ratio (70:30). The I/V curves of the cells having SCGZ electrolyte without and with CuO addition are shown in Figures 6 and 7, respectively. The conductivity and activation energy of conduction are presented in Figure 8. It can be seen that the conductivity increased with increasing operating temperature. The activation energy of conduction (E_a) was 72.34 kJmol^{-1} and 74.93 kJmol^{-1} for SCGZ and SCGZ with CuO, respectively. In this study, adding 0.5 wt% CuO did not significantly affect the conductivity of the electrolyte. The use of CuO as sintering additive in various ceramic electrolytes has been reported differently. It was reported that adding CuO can lead to ionic and electrical properties modification [25,32–36]. Zhang et al. [36] found that addition of 1 wt% CuO improved the sinterability of $Sm_{0.2}Ce_{0.8}O_{1.9}$ (SDC) electrolyte. The SDC could be densified even at lower sintering temperature than 1273 K but the ionic conductivity was also decreased as a result of microstructure alteration. On the other hand, 0.5 mol% of CuO as sintering additive could provide high ionic conductivity and insignificant change in the activation energy of conduction for gadolinium-doped ceria (GDC) electrolyte [37]. In this study, 0.5 wt% CuO was used as a potential sintering additive for SCGZ electrolyte in SOEC, decreasing the sintering temperature without any significant change in the activation energy of conduction.

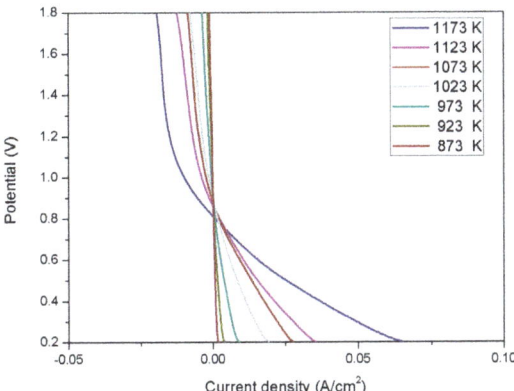

Figure 6. The I/V curves of the SCGZ electrolyte-supported SOEC conducted from 873 K to 1173 K under a constant steam to hydrogen ratio (70:30).

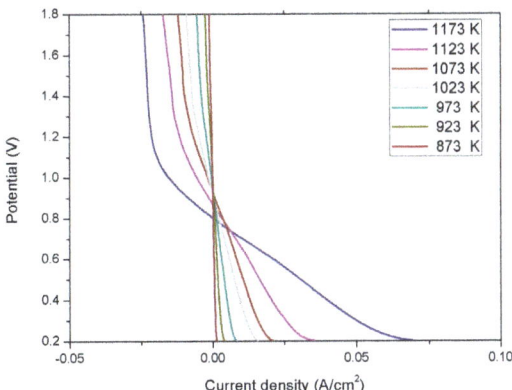

Figure 7. The I/V curves of the SCGZ with 0.5 wt% CuO electrolyte-supported SOEC conducted from 873 K to 1173 K under a constant steam to hydrogen ratio (70:30).

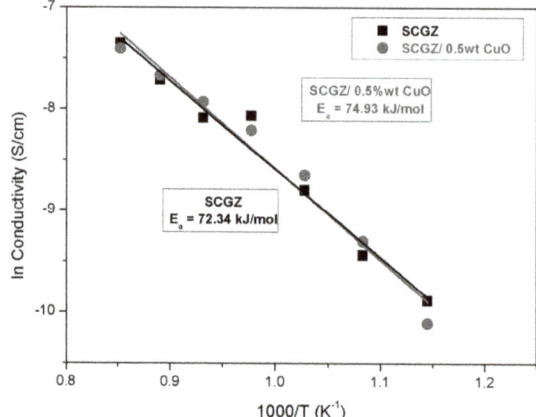

Figure 8. Activation energy of conduction (E_a) for the SOEC having SCGZ electrolyte and 0.5 wt% CuO-added SCGZ electrolyte.

4. Conclusions

0.5 wt% CuO was employed as a sintering additive for SCGZ electrolyte in SOEC. Phase formation, microstructure, relative density, and electrochemical performance of electrolyte-supported SOEC were investigated. Adding 0.5 wt% CuO helped increasing the sinterability of the electrolyte which could achieve 95% relative density with a large grain size at 1573 K. The average grain size was measured at 5.11 µm when sintering temperature was 1573 K for SCGZ with CuO. Phase transformation and impurity was not detected in the electrolyte after adding CuO. Neither peak shifting nor impurity peak were detected in the XRD patterns. Without CuO addition, the SCGZ could not be densified although sintering temperature was increased up to 1673 K. Adding CuO into SCGZ insignificantly affected the electrochemical performance of the cell. The activation energy of conduction (E_a) of the SCGZ with and without CuO was 74.93 kJ mol^{-1} and 72.34 kJ mol^{-1}, respectively.

Author Contributions: Conceptualization, P.K.-L. and R.V.; methodology, P.K.-L. and R.V.; validation, P.K.-L. and R.V.; formal analysis, P.K.-L. and R.V.; investigation, R.V.; resources, P.K.-L.; data curation, P.K.-L. and R.V.; writing—original draft preparation, P.K.-L. and R.V.; writing—review and editing, P.K.-L. and R.V.; visualization, P.K.-L. and R.V.; supervision, P.K.-L., T.S., P.P., T.J., W.N., N.S. and S.P.-O.; project administration, P.K.-L., T.S., P.P. and T.J.; funding acquisition, T.S., P.P., T.J., W.N., N.S. and S.P.-O.

Funding: This research was funded by PTT Innovation Institute, PTT Public Company Limited, Thailand.

Acknowledgments: The acknowledge is made to PTT Innovation Institute, PTT Public Company Limited, Thailand.

Conflicts of Interest: The authors declare no conflict of interest.

References

1. Weng, G.-M.; Vanessa Li, C.-Y.; Chan, K.-Y. Hydrogen battery using neutralization energy. *Nano Energy* **2018**, *53*, 240–244. [CrossRef]
2. Wang, F.-C.; Hsiao, Y.-S.; Yang, Y.-Z. The Optimization of Hybrid Power Systems with Renewable Energy and Hydrogen Generation. *Energies* **2018**, *11*, 1948. [CrossRef]
3. Hawkins, A.S.; McTernan, P.M.; Lian, H.; Kelly, R.M.; Adams, M.W.W. Biological conversion of carbon dioxide and hydrogen into liquid fuels and industrial chemicals. *Curr. Opin. Biotechnol.* **2013**, *24*, 376–384. [CrossRef] [PubMed]
4. Wiesberg, I.L.; de Medeiros, J.L.; Alves, R.M.B.; Coutinho, P.L.A.; Araújo, O.Q.F. Carbon dioxide management by chemical conversion to methanol: Hydrogenation and Bi-Reforming. *Energy Convers. Manag.* **2016**, *125*, 320–335. [CrossRef]

5. Likhittaphon, S.; Panyadee, R.; Fakyam, W.; Charojrochkul, S.; Sornchamni, T.; Laosiripojana, N.; Assabumrungrat, S.; Kim-Lohsoontorn, P. Effect of CuO/ZnO catalyst preparation condition on alcohol-assisted methanol synthesis from carbon dioxide and hydrogen. *Int. J. Hydrogen Energy* **2019**, *44*, 20782–20791. [CrossRef]
6. Javed, H.; Sabato, A.G.; Herbrig, K.; Ferrero, D.; Walter, C.; Salvo, M.; Smeacetto, F. Design and characterization of novel glass-ceramic sealants for solid oxide electrolysis cell (SOEC) applications. *Int. J. Appl. Ceram. Technol.* **2018**, *15*, 999–1010. [CrossRef]
7. Kim-Lohsoontorn, P.; Kim, Y.-M.; Laosiripojana, N.; Bae, J. Gadolinium doped ceria-impregnated nickel–yttria stabilised zirconia cathode for solid oxide electrolysis cell. *Int. J. Hydrogen Energy* **2011**, *36*, 9420–9427. [CrossRef]
8. Chen, T.; Zhou, Y.; Liu, M.; Yuan, C.; Ye, X.; Zhan, Z.; Wang, S. High performance solid oxide electrolysis cell with impregnated electrodes. *Electrochem. Commun.* **2015**, *54*, 23–27. [CrossRef]
9. Kim, J.; Jun, A.; Gwon, O.; Yoo, S.; Liu, M.; Shin, J.; Lim, T.-H.; Kim, G. Hybrid-solid oxide electrolysis cell: A new strategy for efficient hydrogen production. *Nano Energy* **2018**, *44*, 121–126. [CrossRef]
10. Fernández-González, R.; Molina, T.; Savvin, S.; Moreno, R.; Makradi, A.; Núñez, P. Fabrication and electrical characterization of several YSZ tapes for SOFC applications. *Ceram. Int.* **2014**, *40*, 14253–14259. [CrossRef]
11. Talebi, T.; Sarrafi, M.H.; Haji, M.; Raissi, B.; Maghsoudipour, A. Investigation on microstructures of NiO–YSZ composite and Ni–YSZ cermet for SOFCs. *Int. J. Hydrogen Energy* **2010**, *35*, 9440–9447. [CrossRef]
12. Fondard, J.; Bertrand, P.; Billard, A.; Fourcade, S.; Batocchi, P.; Mauvy, F.; Bertrand, G.; Briois, P. Manufacturing and testing of a metal supported Ni-YSZ/YSZ/La_2NiO_4 IT-SOFC synthesized by physical surface deposition processes. *Solid State Ion.* **2017**, *310*, 10–23. [CrossRef]
13. Han, Z.; Yang, Z.; Han, M. Optimization of Ni-YSZ anodes for tubular SOFC by a novel and efficient phase inversion-impregnation approach. *J. Alloys Compd.* **2018**, *750*, 130–138. [CrossRef]
14. Yeh, T.H.; Chiang, C.M.; Lo, W.C.; Chou, C.C. Co-doping effect of divalent (Mg^{2+}, Ca^{2+}, Sr) and trivalent (Y^{3+}) cations with different ionic radii on ionic conductivity of zirconia electrolytes. *J. Chin. Soc. Mech. Eng. Trans. Chin. Inst. Eng. Ser. C* **2009**, *30*, 151–158.
15. Gao, H.; Liu, J.; Chen, H.; Li, S.; He, T.; Ji, Y.; Zhang, J. The effect of Fe doping on the properties of SOFC electrolyte YSZ. *Solid State Ion.* **2008**, *179*, 1620–1624. [CrossRef]
16. Kubrin, R.; Blugan, G.; Kuebler, J. Influence of cerium doping on mechanical properties of tetragonal scandium-stabilized zirconia. *J. Eur. Ceram. Soc.* **2017**, *37*, 1651–1656. [CrossRef]
17. Vijaya Lakshmi, V.; Bauri, R. Phase formation and ionic conductivity studies on ytterbia co-doped scandia stabilized zirconia ($0.9ZrO_2$–$0.09Sc_2O_3$–$0.01Yb_2O_3$) electrolyte for SOFCs. *Solid State Sci.* **2011**, *13*, 1520–1525. [CrossRef]
18. Chen, M.; Gao, H.; Zhang, L.; Xuan, Y.; Ren, J.; Ni, M.; Lin, Z. Unlocking the nature of the co-doping effect on the ionic conductivity of CeO_2-based electrolyte. *Ceram. Int.* **2019**, *45*, 3977–3985. [CrossRef]
19. Abbas, H.A.; Argirusis, C.; Kilo, M.; Wiemhöfer, H.-D.; Hammad, F.F.; Hanafi, Z.M. Preparation and conductivity of ternary scandia-stabilised zirconia. *Solid State Ion.* **2011**, *184*, 6–9. [CrossRef]
20. Shin, H.C.; Yu, J.H.; Lim, K.T.; Lee, H.L.; Baik, K.H. Effects of Partial Substitution of CeO_2 with M_2O_3 (M = Yb, Gd, Sm) on Electrical Degradation of Sc_2O_3 and CeO_2 Co-doped ZrO_2. *J. Korean Ceram. Soc.* **2016**, *53*, 500–505. [CrossRef]
21. Chong, F.D.; Tan, C.Y.; Singh, R.; Muchtar, A.; Somalu, M.R.; Ng, C.K.; Yap, B.K.; Teh, Y.C.; Tan, Y.M. Effect of manganese oxide on the sinterability of 8mol% yttria-stabilized zirconia. *Mater. Charact.* **2016**, *120*, 331–336. [CrossRef]
22. Lee, J.G.; Jeon, O.S.; Ryu, K.H.; Park, M.G.; Min, S.H.; Hyun, S.H.; Shul, Y.G. Effects of 8mol% yttria-stabilized zirconia with copper oxide on solid oxide fuel cell performance. *Ceram. Int.* **2015**, *41*, 7982–7988. [CrossRef]
23. Satardekar, P.; Montinaro, D.; Sglavo, V.M. Fe-doped YSZ electrolyte for the fabrication of metal supported-SOFC by co-sintering. *Ceram. Int.* **2015**, *41*, 9806–9812. [CrossRef]
24. López-Honorato, E.; Dessoliers, M.; Shapiro, I.P.; Wang, X.; Xiao, P. Improvements to the sintering of yttria-stabilized zirconia by the addition of Ni. *Ceram. Int.* **2012**, *38*, 6777–6782. [CrossRef]
25. Myung, J.-H.; Ko, H.J.; Park, H.-G.; Hwan, M.; Hyun, S.-H. Fabrication and characterization of planar-type SOFC unit cells using the tape-casting/lamination/co-firing method. *Int. J. Hydrogen Energy* **2012**, *37*, 498–504. [CrossRef]

26. Likhittaphon, S.; Pukkrueapun, T.; Seeharaj, P.; Wetwathana Hartley, U.; Laosiripojana, N.; Kim-Lohsoontorn, P. Effect of sintering additives on barium cerate based solid oxide electrolysis cell for syngas production from carbon dioxide and steam. *Fuel Process. Technol.* **2018**, *173*, 119–125. [CrossRef]
27. Barsoukov, E.; Mcdonald, J.R. *Impedance Spectroscopy Theory, Experiment, and Applications*; John Wiley & Sons: Hoboken, NJ, USA, 2005.
28. Vu, H.; Nguyen, D.; Fisher, J.G.; Moon, W.-H.; Bae, S.; Park, H.-G.; Park, B.-G. CuO-based sintering aids for low temperature sintering of $BaFe_{12}O_{19}$ ceramics. *J. Asian Ceram. Soc.* **2013**, *1*, 170–177. [CrossRef]
29. Liou, Y.-C.; Wu, C.-T.; Huang, Y.-L.; Chung, T.-C. Effect of CuO on $CaTiO_3$ perovskite ceramics prepared using a direct sintering process. *J. Nucl. Mater.* **2009**, *393*, 492–496. [CrossRef]
30. Amsif, M.; Marrero-López, D.; Ruiz-Morales, J.C.; Savvin, S.N.; Núñez, P. Effect of sintering aids on the conductivity of $BaCe_{0.9}Ln_{0.1}O_3-\delta$. *J. Power Sources* **2011**, *196*, 9154–9163. [CrossRef]
31. Liu, S.-S.; Li, H.; Xiao, W.-D. Sintering effect on crystallite size, hydrogen bond structure and morphology of the silane-derived silicon powders. *Powder Technol.* **2015**, *273*, 40–46. [CrossRef]
32. Presto, S.; Viviani, M. Effect of CuO on microstructure and conductivity of Y-doped $BaCeO_3$. *Solid State Ion.* **2016**, *295*, 111–116. [CrossRef]
33. Zhang, W.; Sun, C. Effects of CuO on the microstructure and electrochemical properties of garnet-type $Li_{6.3}La_3Zr_{1.65}W_{0.35}O_{12}$ solid electrolyte. *J. Phys. Chem. Solids* **2019**, *135*, 109080. [CrossRef]
34. Zhang, C.; Sunarso, J.; Zhu, Z.; Wang, S.; Liu, S. Enhanced oxygen permeability and electronic conductivity of $Ce_{0.8}Gd_{0.2}O_2-\delta$ membrane via the addition of sintering aids. *Solid State Ion.* **2017**, *310*, 121–128. [CrossRef]
35. Nicollet, C.; Waxin, J.; Dupeyron, T.; Flura, A.; Heintz, J.-M.; Ouweltjes, J.P.; Piccardo, P.; Rougier, A.; Grenier, J.-C.; Bassat, J.-M. Gadolinium doped ceria interlayers for Solid Oxide Fuel Cells cathodes: Enhanced reactivity with sintering aids (Li, Cu, Zn), and improved densification by infiltration. *J. Power Sources* **2017**, *372*, 157–165. [CrossRef]
36. Zhang, X.; Decès-Petit, C.; Yick, S.; Robertson, M.; Kesler, O.; Maric, R.; Ghosh, D. A study on sintering aids for $Sm_{0.2}Ce_{0.8}O_{1.9}$ electrolyte. *J. Power Sources* **2006**, *162*, 480–485. [CrossRef]
37. Choi, H.-J.; Na, Y.-H.; Kwak, M.; Kim, T.W.; Seo, D.-W.; Woo, S.-K.; Kim, S.-D. Development of solid oxide cells by co-sintering of GDC diffusion barriers with LSCF air electrode. *Ceram. Int.* **2017**, *43*, 13653–13660. [CrossRef]

© 2019 by the authors. Licensee MDPI, Basel, Switzerland. This article is an open access article distributed under the terms and conditions of the Creative Commons Attribution (CC BY) license (http://creativecommons.org/licenses/by/4.0/).

Article

Comparison of Packed-Bed and Micro-Channel Reactors for Hydrogen Production via Thermochemical Cycles of Water Splitting in the Presence of Ceria-Based Catalysts

Nonchanok Ngoenthong [1], Matthew Hartley [2], Thana Sornchamni [3], Nuchanart Siri-nguan [3], Navadol Laosiripojana [4] and Unalome Wetwatana Hartley [1,4,*]

1. Chemical and Process Engineering, The Sirindhorn International Thai-German Graduate School of Engineering (TGGS), King Mongkut's University of Technology North Bangkok, Bangkok 10800, Thailand; nonchanok.n-cpe2015@tggs-bangkok.org
2. Chemical Engineering Department, Engineering Faculty, King Mongkut's University of Technology North Bangkok, Bangkok 10800, Thailand; matthew.m.hartley@gmail.com
3. PTT Public Company Limited, 555 Vibhavadi Rangsit Road, Chatuchak, Bangkok 10900, Thailand; thana.s@pttplc.com (T.S.); nuchanart.s@pttplc.com (N.S.-n.)
4. Joint Graduate School of Energy and Environment (JGSEE), King Mongkut's University of Technology Thonburi, Bangkok 10140, Thailand; navadol@jgsee.kmutt.ac.th
* Correspondence: unalome.w.cpe@tggs-bangkok.org

Received: 26 August 2019; Accepted: 15 October 2019; Published: 18 October 2019

Abstract: Hydrogen production via two-step thermochemical cycles over fluorite-structure ceria (CeO_2) and ceria-zirconia ($Ce_{0.75}Zr_{0.25}O_2$) materials was studied in packed-bed and micro-channel reactors for comparison purposes. The H_2-temperature program reduction (H_2-TPR) results indicated that the addition of Zr^{4+} enhanced the material's reducibility from 585 µmol/g to 1700 µmol/g, although the reduction temperature increased from 545 to 680 °C. $Ce_{0.75}Zr_{0.25}O_2$ was found to offer higher hydrogen productivity than CeO_2 regardless of the type of reactor. The micro-channel reactor showed better performance than the packed-bed reactor for this reaction.

Keywords: hydrogen production; thermochemical cycles; micro-channel reactor; ceria; ceria-zirconia; water splitting; oxygen carrier

---

1. Introduction

Hydrogen can be utilized in many modern-world applications. Its well-known challenges include production cost, transportation and storage. Hydrogen can be produced by various means, e.g., thermochemical processes, reforming processes, gasification, electrolysis, biological processes, and so on [1–8]. Conventional hydrogen production from either natural gas, coal, or biomass appears to be the most commercially available and affordable, although this unavoidably releases carbon emissions, radioactive elements, and air-borne pollutions into the atmosphere. Hydrogen production from water is a green technology in which water is split, producing high-purity gaseous hydrogen. Recent water splitting processes including, for instance, photo-catalytic, two-step thermochemical cycles, electrolysis and biological processes have been employed to generate high-purity hydrogen. The two-step thermochemical cycle possesses advantages over the others in terms of the product's purity and yield [9]. The two-step thermochemical cycle reaction consists of (1) endothermic reduction, where the metal oxide material is reduced by thermal energy and/or chemical reducing agents, resulting in gaseous oxygen as a by-product and an active and non-stoichiometric reduced metal oxide, (2) exothermic oxidation, where the active metal oxides are oxidized by water, giving high-purity hydrogen as a product while the metal oxide is recycled back into its original stage [10–12]. The operating conditions—i.e., temperature, feed reactant and reaction time—of the two steps are different. Thus, the process can either be carried out in (1) two reactors between which the solid material is moved, or (2) one reactor in which the operating conditions are switched back and forth from the reduction to the oxidation step. The former is suitable for solid materials that have high mechanical strength, have changed phase, and have been fully reduced/oxidized based on their stoichiometry, such as ZnO/Zn [13], CdO/Cd [14], SnO_2/SnO [15], and GeO_2/GeO [16], although the process requires a sophisticated quenching and control system. The latter is commonly used with materials with no phase change. The materials could be either stoichiometric redox materials such as Fe_3O_4/FeO [17], MFe_2O_4 [18], and $CoFe_2O_4/Al_2O_3$ [19], or non-stoichiometric redox materials such as, for example, ceria and ceria-based materials [20–22], and perovskites [23]. Ceria (CeO_2) and ceria-zirconia (CeO_2/ZrO_2) were selected as the oxygen carrier in this work, as they offer (1) high oxygen storage capacity [24,25], (2) high thermal stability, and (3) the possibility of being fabricated as the whole reactor itself. Although the latter process does not require a quenching system and complicated fluidized bed operation, it needs a well-established control system that allows the operational conditions to switch from reduction to oxidation steps precisely. Regardless of the quality of the control system, the switching between conditions (temperature, feed) still causes low overall process efficiency (10 to 50%) [26–29]. Therefore, a reactor that offers rapid heat and mass transfer during the change is beneficial for this process. Additionally, the redox materials must be able to withstand the severe condition swing. This work studied the process under the same reduction and oxidation temperature, aiming to (1) optimize the process efficiency and product yield, (2) ease the control system, and (3) make it possible to use wider variety of the new catalysts. A micro-channel reactor was also applied to enable rapid mas/heat transfer, with shorter residence time.

2. Methodology

2.1. Catalyst Preparation, Substrate Pretreatment and Catalyst Coating

2.1.1. Catalyst Preparation

Nitrate precursors of Ce and Zr were mixed with 0.1 M cetyl trimethyl ammonium bromide (CTAB) at room temperature while NaOH was added dropwise, keeping pH at 11 while stirring. Molar ratio of CTAB to total cations (Ce + Zr) was fixed at 0.8. The mixture was centrifuged at 4000 rpm for 10 min, and then washed with de-ionized water to remove any possible impurities. The precipitate was dried in the oven at 120 °C overnight and calcined at 700 °C for 3 h in air with 2 °C/min heating rate. For the micro-channel reactor, alumina solution was additionally synthesized by hydrolysis of aluminum tri-sec-butoxide (Aldrich, St. Louis, MO, USA, 97%) with a molar ratio of H_2O to Al = 100 to enhance the adhesion force between the catalyst power and the surface of the substrate. The solution was peptized by adding nitric acid (HNO_3/Al = 0.07) and refluxed at 85 °C for 20 h. After that, nitrate precursors of Ce and Zr were dissolved into the prepared alumina solution. The mixed solution was heated at 85 °C until a thixotropic solution was obtained. The gel was dried at 110 °C overnight and calcined at 800 °C for 6 h at a heating rate of 2 °C/min.

2.1.2. Stainless Steel Substrate Preparation

The lab-designed micro-structured stainless steel (316L) plates were commercially made (TGI, Chonburi, Thailand), as shown in Figure 1. Each substrate has 14 channels with 300 μm depth, 370 μm width and 25 mm length per channel. The plates were cleaned by etching with 20% citric acid in an ultrasonic bath for 30 min. They were subsequently annealed at 800 °C for 2 h in air to form a layer of mixed metal oxides for better adhesion strength.

2.1.3. Catalyst Coating

Polyvinyl alcohol was dissolved in deionized water while stirring at 400 rpm at 65 °C for 2 h. The solution was left overnight without stirring. The catalyst powder and acetic acid were added to the solution. The weight ratio of catalyst powder to water to binder to acid was 10 to 84 to 5 to 1. The resulting suspension was stirred at 65 °C for 2 h, cooled to room temperature, and subsequently stirred for 3 days. The non-coating area of the micro-structured stainless steel plates, such as the inlet and the outlet, were covered with the polymer film. The prepared suspension was wash-coated on the micro-channel substrate, then left to dry at room temperature for 6 h. After the removal of the polymer film, the substrate was dried in an oven at 120 °C and calcined at 500 °C for 3 h at a heating rate of 1 °C/min. The 2 micro-channel substrates were then laser-welded together. The gas inlet and outlet at the top and bottom were connected to 1/8-inch stainless steel tubing.

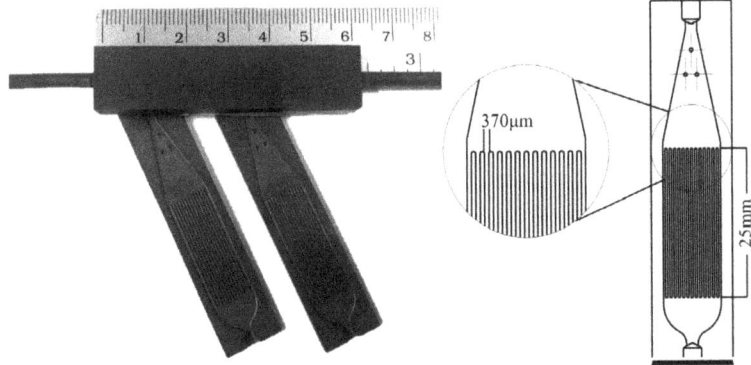

Figure 1. Stainless steel 316L micro-channeled plate.

2.2. Characterization

The crystallinity of the synthesized catalysts was characterized using X-ray diffraction (XRD). The morphology of the stainless steel substrates, before and after the annealing process at different temperatures, was characterized using scanning electron microscopy (SEM). The reduction temperature of the samples was investigated using the H_2-temperature program reduction technique (H_2-TPR). During the H_2-TPR, 10% H_2/Ar was passed through the catalyst's bed in the reactor. The temperature was increased from room temperature to 950 °C, at a heating rate of 5 °C/min, and held for 30 min. The gaseous products were analyzed using on-line mass spectrometer (MS, GSD 320 O1, OmniStar gas analysis) for all experiments.

2.3. Experimental Set-Up

The catalysts were pelletized and sieved to 180–212 μm for the packed-bed reactor, and <38 μm for the micro-channel reactor. The weight of the catalyst was 1 g and 8–15 mg for the packed-bed reactor and the micro-channel reactor, respectively. Each catalyst was placed inside the quartz tube packed-bed reactor (i.d. = 10 mm, length = 50 cm) between two layers of quartz wool. The packed-bed and micro-channel reactors were placed in the middle of an electrical furnace (Inconel, 20 cm heating zone). The schematic diagram of the experiments is shown in Figure 2. The system was purged by 300 mL/min of Ar using a mass flow controller (New Flow-TLFC-00-A-1-W-2, 10–500 mL/min) at room temperature for 1 h. Each catalyst was reduced before use in 10% H_2/Ar at 700 and 900 °C, in accordance with the TPR results, for 30 min. Steam was generated using a steam generator, whereby the amount of water was controlled by peristaltic pump (BT100M Model, 0.00067–65 mL/min). Steam was delivered to the reactor system through a 170 °C trace-heated line to prevent condensation. The total flow rate of 200 mL/min, which ensured the reaction control regime, was fixed for all of the experiments.

The reaction was isothermally operated at 700 and 900 °C. The gaseous product stream was analyzed using an on-line mass spectrometer (Quadrupole, Omnistar, GSD 320 O1 Model).

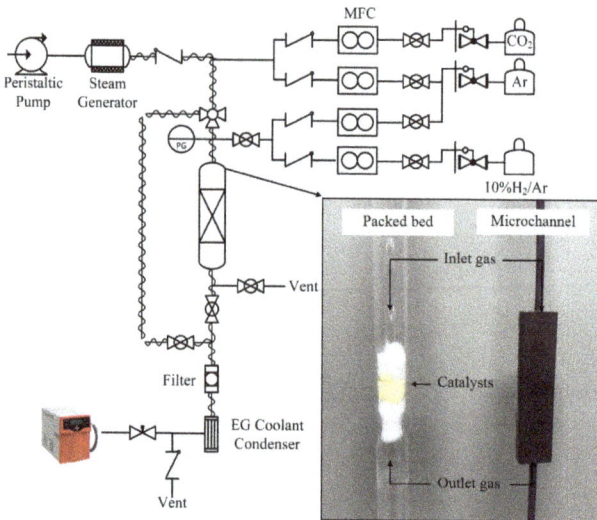

Figure 2. Rig schematic diagram for packed-bed and micro-channel process.

3. Result and Discussion

3.1. Characterization

Figure 3 shows diffractograms of all of the prepared catalysts, compared with pure alumina, shown as (a) CeO_2, (b) $Ce_{0.75}Zr_{0.25}O_2$, (c) 10%CeO_2/Al_2O_3, (d) 10%$Ce_{0.75}Zr_{0.25}O_2/Al_2O_3$, and (e) pure alumina.

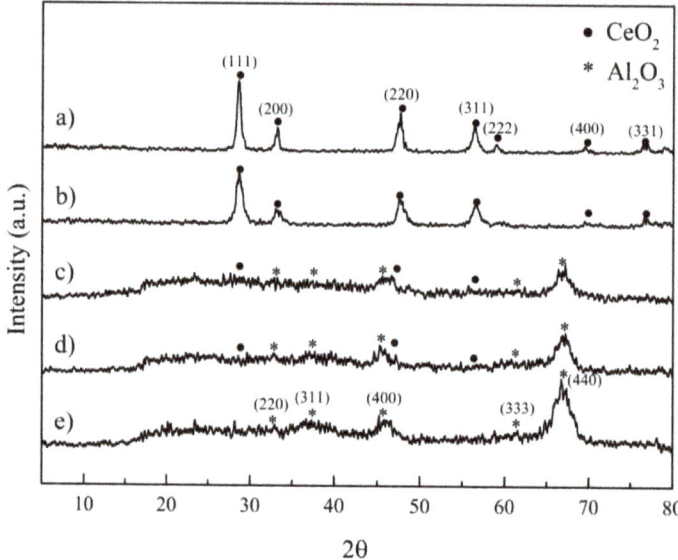

Figure 3. XRD patterns of (a) CeO_2, (b) $Ce_{0.75}Zr_{0.25}O_2$, (c) 10% CeO_2/Al_2O_3, (d) 10% $Ce_{0.75}Zr_{0.25}O_2/Al_2O_3$, and (e) pure alumina.

CeO$_2$ and Ce$_{0.75}$Zr$_{0.25}$O$_2$ possessed a fluorite cubic structure with a face-centered cubic crystal system (FCC), which offers a high thermal stability and oxygen exchange kinetic rate [30–32]. CeO$_2$ and Ce$_{0.75}$Zr$_{0.25}$O$_2$ presented a lattice plane corresponding to the (111), (200), (220), (311), (222), (400) and (331) [33,34]. Their average crystal size, calculated using the Scherer equation, was 15.15 nm and 9.01 nm, respectively. Pure alumina was η-alumina and had a bayerite structure. Each peak corresponded to the (220), (311), (400), (333), and (440) lattice planes [35,36]. Both 10% CeO$_2$/Al$_2$O$_3$ and 10% Ce$_{0.75}$Zr$_{0.25}$O$_2$/Al$_2$O$_3$ showed η-alumina bayerite structures as a major crystalline phase, while CeO$_2$ and Ce$_{0.75}$Zr$_{0.25}$O$_2$ showed them as minor crystalline phase. The non-alumina catalysts were selected for the further study.

CeO$_2$ and Ce$_{0.75}$Zr$_{0.25}$O$_2$ showed two main reduction peaks at different temperatures, shown in Figure 4. Ceria and ceria-zirconia started to be reduced at the same temperatures, with the first peak at 300 °C (peaks α) and the second peak at 650 °C (peaks β). However, broader reduction peaks were observed in the ceria-zirconia due to the larger amount of oxygen release in ceria-zirconia. This led to a higher average reduction temperature for ceria-zirconia (680 and 950 °C), when compared to ceria (545 and 900°C), respectively. The results agreed with previous works reported by other researchers [37–40]. The first peaks of both catalysts were defined as surface reduction, evidenced by the steep reduction peaks, while the second peaks comprised bulk reductions, which were much broader compared to the first peaks due to the much slower solid-state oxygen diffusion within the materials. From the H$_2$-TPR profiles of both catalysts, it can be seen that ceria-zirconia had a higher reduction rate than ceria, and released a higher amount of oxygen within the studied temperature range. The addition of Zr into the ceria catalyst system has been suggested to affect the material's cell volume, resulting in an increase in surface area [41–44]. The degree of the reduction, represented by the non-stoichiometric oxygen release (δ), is calculated and tabulated in Table 1. For CeO$_2$, the surface oxygen was released at 545 °C, giving δ at 0.046 (theoretically maximum at 0.5) and percentage of reduction at 9.34%, where the material was reduced to CeO$_{1.95}$. The second peak of CeO$_2$ showed that the bulk oxygen was reduced at 900 °C, giving 0.074 of δ, which is equal to a reduction of 14.64%. At this stage, the CeO$_{1.88}$ became CeO$_{1.94}$. Similarly, ceria-zirconia was surface reduced and bulk reduced at 680 °C and 950 °C, respectively. The first reduction peak represented a δ of 0.277, with a reduction degree of 54.40%, while the second peak gave a δ of 0.222, with a reduction degree of 44.46%. The non-stoichiometric molecular formula of the ceria-zirconia after being reduced at 950 °C was Ce$_{0.75}$Zr$_{0.25}$O$_{1.51}$. Ce$_{0.75}$Zr$_{0.25}$O$_2$ conclusively showed better performance, compared to ceria, in terms of reduction rate and reducibility.

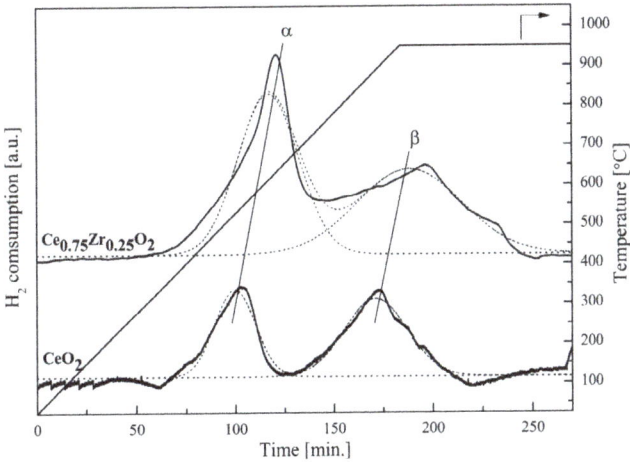

Figure 4. TPR profiles of CeO$_2$ and Ce$_{0.75}$Zr$_{0.25}$O$_2$ for temperatures ranging from 30 to 950 °C, at heating rate of 5 °C/min, using 10% H$_2$/Ar.

Table 1. H_2-TPR results of CeO_2 and $Ce_{0.75}Zr_{0.25}O_2$

Catalysts	Peak α			Peak β			Total OSC (μmol/g)	Total % Reduction (δ)
	T_{red} (°C)	OSC (μmol/g)	% Reduction (δ)	T_{red} (°C)	OSC (μmol/g)	% Reduction (δ)		
$CeO_{2-\delta}$	545	585	9.34% (δ = 0.046)	900	915	14.64% (δ = 0.074)	1500	23.98% (δ = 0.12)
$Ce_{0.75}Zr_{0.25}O_{2-\delta}$	680	1700	547.40% (δ = 0.272)	950	1390	44346% (δ = 0.222)	3090	98.86% (δ = 0.49)

Where OSC stands for oxygen storage capacity, calculated from H_2 consumption. The percentage of reduction degree is calculated using Equations (1)–(3) below:

$$\%X_{red} = \left(\frac{n_{[O]real}}{n_{[O]max}}\right) \times 100 \qquad (1)$$

$$n_{[O]real} = \frac{n_{H_2,consumed}}{m_{solid}} \qquad (2)$$

$$n_{[O]max} = \left(\frac{m_{solid}}{MW_{solid}}\right) \times \delta_{max} \qquad (3)$$

where $n_{[O]max}$ is the maximum amount of O_2 release/uptake (mol/g) as a function of δ, δ is the stoichiometric coefficient of O in lattice (for this material, δ is 0.5 for maximum O release/uptake), $n_{[O]real}$ is the number mole of O released per gram of the catalyst, which is equal to an integration of the area under the H_2 consumption curve, m_{solid} is the weight of the solid reactant, and MW_{solid} is the molecular weight of the catalyst [45].

Figure 5 shows the surface morphography of the prepared substrates (a) before and after annealing process at (b) 600, (c) 700, and (d) 800 °C. From the results, it can be seen that the oxides of the stainless steel's surface were formed by annealing, occurring best at the highest temperature: 800 °C.

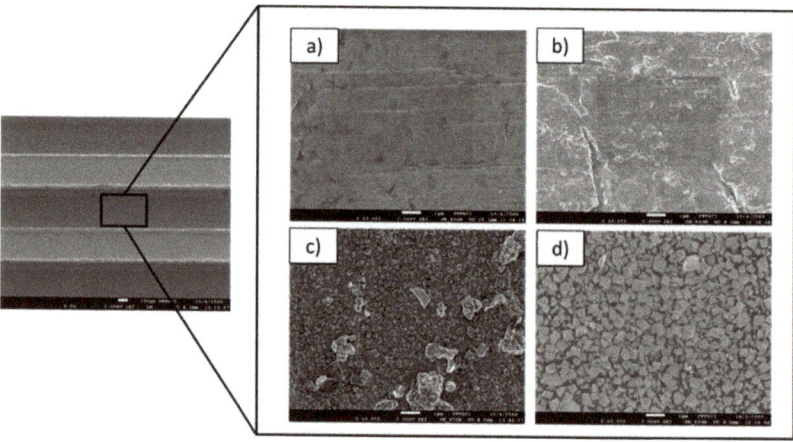

Figure 5. Morphology of the substrate surface analyzed using SEM technique: (a) before annealing and after annealing process at (b) 600, (c) 700, and (d) 800 °C.

Energy-dispersive X-ray spectroscopy technique (EDX) was applied to identify the oxides which were formed with metals consisting of the stainless steel. Table 2 presents the percentage of each element. The results suggested that the formation of oxides increased with an increase in annealing temperature, as evidenced by the higher percentage of oxygen. The results are in agreement with the SEM results, as reported in the previous section.

Table 2. Percentage of each element on the surface of the substrate.

Condition	O	Cr	Mn	Fe	Ni	Others	Total
Before annealing	7.69	28.61	–	47.87	5.65	10.18	100
Annealed at 600 °C	24.84	29.53	2.72	35.95	4.22	2.75	100
Annealed at 700 °C	30.04	26.34	–	30.44	3.75	9.43	100
Annealed at 800 °C	41.78	27.90	4.82	18.70	–	6.79	100

3.2. Catalytic Performance Experiments

3.2.1. Effect of ZrO$_2$ Addition

Catalytic performances of CeO$_2$ and Ce$_{0.75}$Zr$_{0.25}$O$_2$ towards the water splitting were tested in a packed bed reactor. The oxidation and reduction temperatures were paired at the same temperature to avoid temperature switching, aiming to maximize the overall process efficiency and minimize the thermal stress of the reactor. The selected temperatures of reduction and oxidation (Tred/Tox) were 700/700 and 900/900 °C.

From Figure 6, it can be seen that Ce$_{0.75}$Zr$_{0.25}$O$_2$ had better performance towards this reaction than CeO$_2$ for both selected temperatures. The effect was more obvious at the lower temperature (700 °C) than the higher temperature (900 °C). At 700 °C, the H$_2$ production of Ce$_{0.75}$Zr$_{0.25}$O$_2$ was 88.71% more than that produced via CeO$_2$ (increased from 483.42 to 912.26 μmol/g), although it was only 8.40% higher (increased from 1563.21 to 1694.55 μmol/g) at 900 °C. This result shared the same trend as previous work reported by Z. Zhao et al. [46], although both catalysts in this research offered around two times higher of H$_2$ productivity for both temperatures, which could be due to the benefit of a surfactant-assisted method which allows smaller fine particles and larger specific surface area [47].

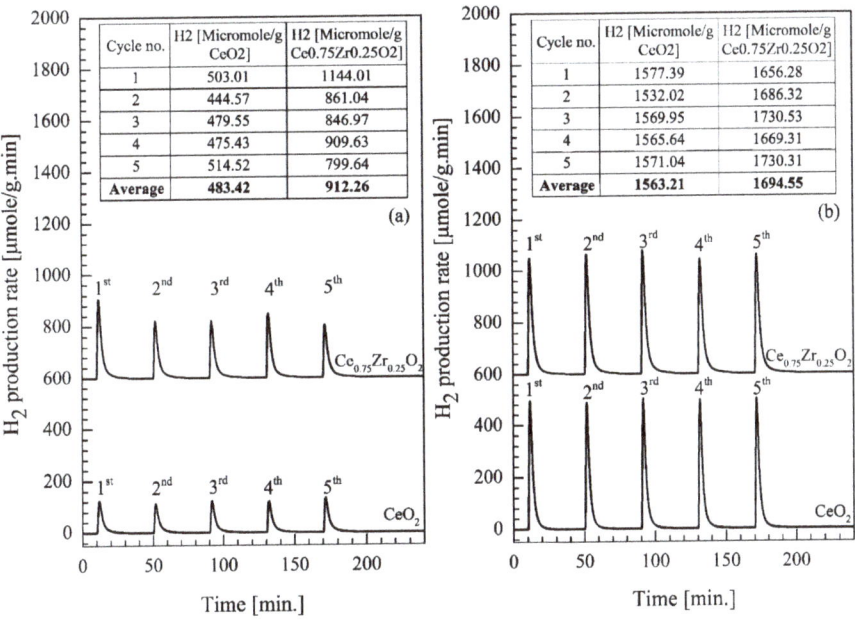

Figure 6. H$_2$ productivity over CeO$_2$ and Ce$_{0.75}$Zr$_{0.25}$O$_2$ in a packed-bed reactor, at the same reduction/oxidation temperature (a) 700/700 °C and (b) 900/900 °C.

3.2.2. Comparison of Micro-Channel Reactor and Packed-Bed Reactor

$Ce_{0.75}Zr_{0.25}O_2$ was selected for this experiment. H_2 production via both packed-bed and micro-channel reactors were compared. Uncoated micro-channel reactor was also introduced to the reactant stream and performed as a blank test. H_2 productivity in all experiments was calculated per weight of catalyst used. H_2 productivity from the blank micro-channel reactor, uncoated reactor, was subtracted from the catalyst-coated micro-channel reactor. The blank test was performed only for the micro-channel reactor, because the formation of metal oxides on the substrate surface after annealing, such as on Cr_2O_3, Mn_2O_3 and Fe_2O_3, could possibly be involved in the catalytic reaction. From Figure 7, the 5-cycle average amount of H_2 production using the packed-bed reactor and the micro-channel reactor was 912.26 and 14,308.32 µmol/g. Thus, micro-channel reactor showed roughly 16 times better performance than the packed-bed reactor, in terms of H_2 production. This was suggested to be the effect of its high surface-to-volume ratio, leading to an intrinsic reaction occurring at the molecular level [48–51]. It can be noticed that the H_2 productivity of the packed-bed reactor decreased while that of the micro-channel reactor increased when the number of cycles increased. The decrease in H_2 productivity in the packed-bed reactor was presumably due to the catalyst's coagulation when repetitively used at such temperatures. On the other hand, the increase in H_2 productivity in the micro-channel reactor was possibly the result of the reactive oxides, in which they were formed by the reaction of metals in the stainless steel and the oxygen in the system. Thus, the more cycles the reaction was run, the more H_2 productivity was achieved. However, these H_2 productivities in both type of reactor were supposed to be constant after a certain number of cycles.

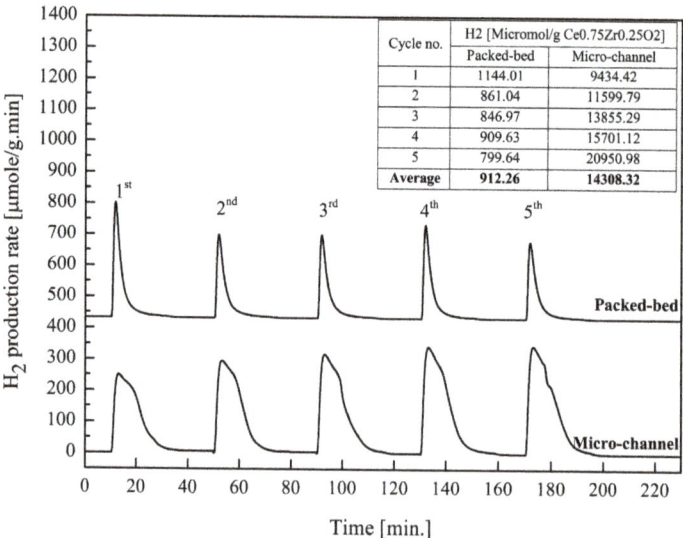

Figure 7. Comparison of H_2 productivity over $Ce_{0.75}Zr_{0.25}O_2$ received from the micro-channel reactor and the packed-bed reactor at reduction/oxidation temperature at 700/700 °C.

In addition, the decline of the H_2 production rate in the micro-channel reactor was noticed to be much slower than that in the packed-bed reactor, leading to a longer reaction time. This was due to (1) the micro-channel reactor having a higher active surface area to volume, allowing better access of the reactants to the catalyst's surface, and (2) channeling and/or mass transfer limitation occurring in the packed-bed reactor, and thus, the reaction time being shorter than it should be.

3.2.3. The Influence of Alumina Addition into the Catalyst System in the Micro-Channel Reactor

Alumina was added to the $Ce_{0.75}Zr_{0.25}O_2$ catalyst system and used in this experiment, as it was believed to increase the adhesion force between the active catalyst powders and the surface of stainless steel substrates [52,53]. From Figure 8, it can be seen that the 5-cycle average H_2 amount, produced in the presence of the bare reactor, $Ce_{0.75}Zr_{0.25}O_2$, and $Ce_{0.75}Zr_{0.25}O_2/Al_2O_3$, was estimated at 130.83, 245.61 and 108.45 µmol, respectively. This concluded that the addition of alumina had a negative effect on H_2 productivity. In addition, H_2 production when using $Ce_{0.75}Zr_{0.25}O_2/Al_2O_3$ was lower than when using the blank reactor. This means that the addition of alumina inhibited the access of water to react with $Ce_{0.75}Zr_{0.25}O_2$. Meanwhile, the surface of the blank reactor was obviously a catalyst itself, as it was formed by the oxidation of metals in stainless steel, such as Cr, Mn, and Fe. These metal oxides are known as redox catalysts, and could therefore catalyze this reaction.

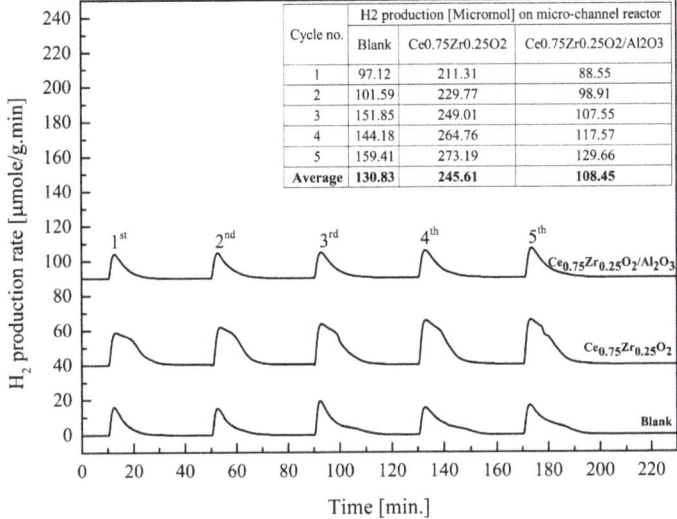

Figure 8. H_2 productivity over $Ce_{0.75}Zr_{0.25}O_2$ and $Ce_{0.75}Zr_{0.25}O_2/Al_2O_3$ coated in a micro-channel reactor, compared to a blank micro-channel reactor at the same reduction/oxidation temperature 700/700 °C.

4. Conclusions

$Ce_{0.75}Zr_{0.25}O_2$ was found to have better catalytic performance towards the two-step thermochemical cycles of water splitting, compared to CeO_2. The higher oxygen storage capacity was suggested to be the cause. Oxygen mobility in the lattice depended on the effective radius of the cations. Thus, the partial substitution of Ce^{4+} (higher ionic radius of 0.97 Å) with Zr^{4+} (smaller ionic radius of 0.84 Å) could create a smaller unit cell volume and larger channel radius in the lattice while the desired fluorite structure of ceria was remained. The infinitesimal cell volume required less energy for the hopping of oxygen ions; therefore, the active oxygen could be easily migrated from one vacancy to the others through the channel radius in the lattice. In a packed-bed reactor, it gave 1694 µmol/g of H_2 productivity at 900 °C. Because the micro-channel reactor was fabricated from stainless steel, the highest operating temperature of the micro-channel reactor was limited to 700 °C. The micro-channel reactor was proved to show 16 times higher H_2 productivity compared with the packed-bed reactor when operated at 700 °C. This was the result of the high surface-to-volume ratio of the micro-channel reactor, which allows better access of the gaseous reactant to react with the catalyst.

Author Contributions: Conceptualization, U.W.H. and T.S.; Methodology, N.N.; Validation, U.W.H., N.S.-n. and N.N.; Formal Analysis, M.H.; Investigation, N.L.; Data Curation, N.N.; Writing-Original Draft Preparation, N.N.; Writing-Review & Editing, M.H.; Visualization, U.W.H.; Supervision, N.L.; Project Administration, N.S.-n.; Funding Acquisition, T.S.

Funding: This research was funded by PTT, Thailand Research Fund [RSA6180061] and National Research Council of Thailand [KMUTNB-GOV-58-47, KMUTNB-GOV-59-43, KMUTNB-GOV-60-55, KMUTNB-61-GOV-03-44/KMUTNB-61-GOV-C1-46].

Acknowledgments: PTT, Thailand Research Fund (RSA6180061) and National Research Council of Thailand (KMUTNB-GOV-58-47, KMUTNB-GOV-59-43, KMUTNB-GOV-60-55, KMUTNB-61-GOV-03-44/KMUTNB-61-GOV-C1-46) are acknowledged for financial support.

Conflicts of Interest: The authors declare no conflict of interest.

References

1. Balat, M. Potential importance of hydrogen as a future solution to environmental and transportation problems. *Int. J. Hydrog. Energy* **2008**, *33*, 4013–4029. [CrossRef]
2. Mathur, H.B.; Das, L.M.; Patro, T.N. Hydrogen fuel utilization in CI engine powered end utility systems. *Int. J. Hydrog. Energy* **1992**, *17*, 369–374. [CrossRef]
3. Chang, A.C.C.; Chang, H.F.; Lin, F.J.; Lin, K.H.; Chen, C.H. Biomass gasification for hydrogen production. *Int. J. Hydrog. Energy* **2011**, *36*, 14252–14260. [CrossRef]
4. Appleby, A.J. Fuel cell technology and innovation. *J. Power Sources* **1992**, *37*, 223–239. [CrossRef]
5. Ismagilov, Z.R.; Matus, E.V.; Ismagilov, I.Z. Hydrogen production through hydrocarbon fuel reforming processes over Ni based catalysts. *Catal. Today* **2019**, 166–182. [CrossRef]
6. Lecart, B.; Devalette, M.; Manaud, J.P.; Meunier, G.; Hagenmuller, P. A new thermochemical process for hydrogen production. *Int. J. Hydrog. Energy* **1979**, *4*, 7–11. [CrossRef]
7. DeLuchi, M.A. Hydrogen vehicles: An evaluation of fuel storage, performance, safety, environmental impacts, and cost. *Int. J. Hydrog. Energy* **1989**, *14*, 81–130. [CrossRef]
8. Nikolaidis, P.; Poullikkas, A. A comparative overview of hydrogen production processes. *Renew. Sustain. Energy Rev.* **2017**, *67*, 597–611. [CrossRef]
9. Agrafiotis, C.; Roeb, M.; Konstandopoulos, A.G. Solar water splitting for hydrogen production with monolithic reactors. *Sol. Energy* **2005**, *79*, 409–421. [CrossRef]
10. Dersch, J.; Mathijsen, A.; Roeb, M.; Sattler, C. Modelling of a solar thermal reactor for hydrogen generation. In Proceedings of the 5th International Modelica Conference, Vienna, Austria, 4–5 September 2006; pp. 441–448. Available online: http://elib.dlr.de/46853/ (accessed on 15 June 2019).
11. Venstrom, L.J.; Petkovich, N.; Rudisill, S.; Stein, A.; Davidson, J.H. The effects of morphology on the oxidation of ceria by water and carbon dioxide. *J. Sol. Energy Eng.* **2012**, *134*, 011005. [CrossRef]
12. Roeb, M.; Sattler, C.; Klüser, R. Solar hydrogen production by a two-step cycle based on mixed iron oxides. In Proceedings of the ASME 2005 International Solar Energy Conference, Orlando, FL, USA, 6–12 August 2005; pp. 671–678. [CrossRef]
13. Steinfeld, A. Solar hydrogen production via a two-step water-splitting thermochemical cycle based on Zn/ZnO redox reactions. *Int. J. Hydrog. Energy* **2002**, *27*, 611–619. [CrossRef]
14. Sibieude, F.; Ducarroir, M.; Tofighi, A.; Ambriz, J. High temperature experiments with a solar furnace: The decomposition of Fe3O4, Mn3O4, CdO. *Int. J. Hydrog. Energy* **1982**, *7*, 79–88. [CrossRef]
15. Popa, S.G.; Ungureanu, B.S.; Gheonea, I.A. Experimental Study of $SnO_2/SnO/Sn$ Thermochemical Systems for Solar Production of Hydrogen. *Rom. J. Morphol Embryol.* **2015**, *56*, 1495–1502. [CrossRef] [PubMed]
16. Kang, K.S.; Kim, C.H.; Cho, W.C.; Bae, K.K.; Kim, S.H.; Park, C.S. Novel two-step thermochemical cycle for hydrogen production from water using germanium oxide: KIER 4 thermochemical cycle. *Int. J. Hydrog. Energy* **2009**, *34*, 4283–4290. [CrossRef]
17. Charvin, P.; Abanades, S.; Flamant, G.; Lemort, F. Two-step water splitting thermochemical cycle based on iron oxide redox pair for solar hydrogen production. *Energy* **2007**, *32*, 1124–1133. [CrossRef]
18. Tamaura, Y.; Ueda, Y.; Matsunami, J. Solar hydrogen production by using ferrites. *Sol. Energy* **1999**, *65*, 55–57. [CrossRef]
19. Scheffe, J.R.; Li, J.; Weimer, A.W. A spinel ferrite/hercynite water-splitting redox cycle. *Int. J. Hydrog. Energy* **2010**, *35*, 3333–3340. [CrossRef]

20. Chueh, W.C.; Haile, S.M. A thermochemical study of ceria: Exploiting an old material for new modes of energy conversion and CO_2 mitigation. *Philos. Trans. R. Soc. A Math. Phys. Eng. Sci.* **2010**, *368*, 3269–3294. [CrossRef]
21. Meng, Q.L.; Lee, C.I.l.; Ishihara, T.; Kaneko, H.; Tamaura, Y. Reactivity of CeO_2-based ceramics for solar hydrogen production via a two-step water-splitting cycle with concentrated solar energy. *Int. J. Hydrog. Energy* **2011**, *36*, 13435–13441. [CrossRef]
22. Furler, P.; Scheffe, J.; Gorbar, M.; Moes, L.; Vogt, U.; Steinfeld, A. Solar thermochemical CO_2 splitting utilizing a reticulated porous ceria redox system. *Energy Fuels* **2012**, *26*, 7051–7059. [CrossRef]
23. Scheffe, J.R.; Weibel, D.; Steinfeld, A. Lanthanum-strontium-manganese perovskites as redox materials for solar thermochemical splitting of H_2O and CO_2. *Energy Fuels* **2013**, *27*, 4250–4257. [CrossRef]
24. Mamontov, E.; Egami, T.; Brezny, R.; Koranne, M.; Tyagi, S. Lattice defects and oxygen storage capacity of nanocrystalline ceria and ceria-zirconia. *J. Phys. Chem. B* **2000**, *104*, 11110–11116. [CrossRef]
25. Epifani, M.; Andreu, T.; Abdollahzadeh-Ghom, S.; Arbiol, J.; Morante, J.R. Synthesis of ceria-zirconia nanocrystals with improved microstructural homogeneity and oxygen storage capacity by hydrolytic sol-gel process in coordinating environment. *Adv. Funct. Mater.* **2012**, *22*, 2867–2875. [CrossRef]
26. Bader, R.; Venstrom, L.J.; Davidson, J.H.; Lipiński, W. Thermodynamic analysis of isothermal redox cycling of ceria for solar fuel production. *Energy Fuels* **2013**, *27*, 5533–5544. [CrossRef]
27. Fueki, K. Efficiency of thermochemical production of hydrogen. *Int. J. Hydrog. Energy* **1976**, *1*, 129–131. [CrossRef]
28. Scheffe, J.R.; Steinfeld, A. Thermodynamic analysis of cerium-based oxides for solar thermochemical fuel production. *Energy Fuels* **2012**, *26*, 1928–1936. [CrossRef]
29. Falter, C.; Pitz-Paal, R. Energy analysis of solar thermochemical fuel production pathway with a focus on waste heat recuperation and vacuum generation. *Sol. Energy* **2018**, *176*, 230–240. [CrossRef]
30. Pengpanich, S.; Meeyoo, V.; Rirksomboon, T.; Bunyakiat, K. Catalytic oxidation of methane over CeO_2-ZrO_2 mixed oxide solid solution catalysts prepared via urea hydrolysis. *Appl. Catal. A Gen.* **2002**, *234*, 221–233. [CrossRef]
31. Li, J.; Liu, X.; Zhan, W.; Guo, Y.; Guo, Y.; Lu, G. Preparation of high oxygen storage capacity and thermally stable ceria-zirconia solid solution. *Catal. Sci. Technol.* **2016**, *6*, 897–907. [CrossRef]
32. Reddy, B.M.; Reddy, G.K.; Reddy, L.H.; Ganesh, I. Synthesis of Nanosized Ceria-Zirconia Solid Solutions by a Rapid Microwave-Assisted Combustion Method. *Open Phys. Chem. J.* **2009**, *3*, 24–29. [CrossRef]
33. Sujana, M.G.; Chattopadyay, K.K.; Anand, S. Characterization and optical properties of nano-ceria synthesized by surfactant-mediated precipitation technique in mixed solvent system. *Appl. Surf. Sci.* **2008**, *254*, 7405–7409. [CrossRef]
34. Shih, C.J.; Chen, Y.J.; Hon, M.H. Synthesis and crystal kinetics of cerium oxide nanocrystallites prepared by co-precipitation process. *Mater. Chem. Phys.* **2010**, *121*, 99–102. [CrossRef]
35. Osman, A.I.; Abu-Dahrieh, J.K.; Rooney, D.W.; Halawy, S.A.; Mohamed, M.A.; Abdelkader, A. Effect of precursor on the performance of alumina for the dehydration of methanol to dimethyl ether. *Appl. Catal. B Environ.* **2012**, *127*, 307–315. [CrossRef]
36. Takashi, S.; Hideo, W.; Masayoshi, F.; Minoru, T. Structural Properties and Surface Characteristics on Aluminum Oxide Powders. *Rev. Med. Chile* **2009**, *9*, 23–31.
37. Wang, X.; Liu, D.; Li, J.; Zhen, J.; Zhang, H. Clean synthesis of Cu_2O@CeO_2 core@shell nanocubes with highly active interface. *NPG Asia Mater.* **2015**, *7*, 158–164. [CrossRef]
38. Zhang, X.M.; Deng, Y.Q.; Tian, P.; Shang, H.H.; Xu, J.; Han, Y.F. Dynamic active sites over binary oxide catalysts: In situ/operando spectroscopic study of low-temperature CO oxidation over MnO_x-CeO_2 catalysts. *Appl. Catal. B Environ.* **2016**, *191*, 179–191. [CrossRef]
39. Liu, L.; Shi, J.; Zhang, X.; Liu, J. Flower-Like Mn-Doped CeOMicrostructures: Synthesis, Characterizations, and Catalytic Properties. *J. Chem.* **2015**, *2015*, 254750. [CrossRef]
40. MacIel, C.G.; Silva, T.D.F.; Hirooka, M.I.; Belgacem, M.N.; Assaf, J.M. Effect of nature of ceria support in CuO/CeO_2 catalyst for PROX-CO reaction. *Fuel* **2012**, *97*, 245–252. [CrossRef]
41. Biswas, P.; Kunzru, D. Steam reforming of ethanol for production of hydrogen over Ni/CeO_2-ZrO_2 catalyst: Effect of support and metal loading. *Int. J. Hydrog. Energy* **2007**, *32*, 969–980. [CrossRef]
42. Kim, D.J. Lattice Parameters, Ionic Conductivities, and Solubility Limits in Fluorite-Structure MO_2 Oxide [M = Hf^{4+}, Zr^{4+}, Ce^{4+}, Th^{4+}, U^{4+}] Solid Solutions. *J. Am. Ceram. Soc.* **1989**, *72*, 1415–1421. [CrossRef]

43. Kang, K.; Kim, C.; Park, C.; Kim, J. Hydrogen Reduction and Subsequent Water Splitting. *J. Ind. Eng. Chem.* **2007**, *13*, 657–663.
44. Córdoba, L.F.; Martínez-Hernández, A. Preferential oxidation of CO in excess of hydrogen over Au/CeO$_2$-ZrO$_2$ catalysts. *Int. J. Hydrog. Energy* **2015**, *40*, 16192–16201. [CrossRef]
45. Le Gal, A.; Abanades, S.; Flamant, G. CO$_2$ and H$_2$O splitting for thermochemical production of solar fuels using nonstoichiometric ceria and ceria/zirconia solid solutions. *Energy Fuels* **2011**, *25*, 4836–4845. [CrossRef]
46. Zhao, Z.; Uddi, M.; Tsvetkov, N.; Yildiz, B.; Ghoniem, A.F. Enhanced intermediateerature CO$_2$ splitting using nonstoichiometric ceria and ceria-zirconia. *Phys. Chem. Chem. Phys.* **2017**, *19*, 25774–25785. [CrossRef] [PubMed]
47. Sukonket, T.; Khan, A.; Saha, B. Influence of the Catalyst Preparation Method, Surfactant Amount, and Steam on CO$_2$ Reforming of CH4 over 5Ni/Ce0.6Zr 0.4O$_2$ Catalysts. *Energy Fuels* **2011**, *25*, 864–877. [CrossRef]
48. Kolb, G. Review: Microstructured reactors for distributed and renewable production of fuels and electrical energy. *Chem. Eng. Process. Process Intensif.* **2013**, *65*, 1–44. [CrossRef]
49. Pennemann, H.; Watts, P.; Haswell, S.J.; Hessel, V.; Löwe, H. Benchmarking of microreactor applications. *Org. Process Res. Dev.* **2004**, *8*, 422–439. [CrossRef]
50. Mathieu-Potvin, F.; Gosselin, L.; Da Silva, A.K. Optimal geometry of catalytic microreactors: Maximal reaction rate density with fixed amount of catalyst and pressure drop. *Chem. Eng. Sci.* **2012**, *73*, 249–260. [CrossRef]
51. Lau, W.N.; Yeung, K.L.; Martin-Aranda, R. Knoevenagel condensation reaction between benzaldehyde and ethyl acetoacetate in microreactor and membrane microreactor. *Microporous Mesoporous Mater.* **2008**, *115*, 156–163. [CrossRef]
52. Peela, N.R.; Mubayi, A.; Kunzru, D. Washcoating of γ-alumina on stainless steel microchannels. *Catal. Today* **2009**, *147*, 17–23. [CrossRef]
53. Wu, X.; Weng, D.; Zhao, S.; Chen, W. Influence of an aluminized intermediate layer on the adhesion of a γ-Al$_2$O$_3$ washcoat on FeCrAl. *Surf. Coat. Technol.* **2005**, *190*, 434–439. [CrossRef]

© 2019 by the authors. Licensee MDPI, Basel, Switzerland. This article is an open access article distributed under the terms and conditions of the Creative Commons Attribution (CC BY) license (http://creativecommons.org/licenses/by/4.0/).

Article

Economic Viability and Environmental Efficiency Analysis of Hydrogen Production Processes for the Decarbonization of Energy Systems

Li Xu [1,2], Ying Wang [1], Syed Ahsan Ali Shah [1,*], Hashim Zameer [1], Yasir Ahmed Solangi [1], Gordhan Das Walasai [3] and Zafar Ali Siyal [4]

1. College of Economics and Management, Nanjing University of Aeronautics and Astronautics, 29 Jiangsu Avenue, Nanjing 211106, China
2. College of Finance, Jiangsu Vocational Institute of Commerce, Nanjing 211168, China
3. Department of Mechanical Engineering, Quaid-e-Awam University of Engineering, Science and Technology, Nawabshah 67480, Pakistan
4. Department of Energy and Environment, Quaid-e-Awam University of Engineering, Science and Technology, Nawabshah 67480, Pakistan
* Correspondence: ahsan.shah@nuaa.edu.cn; Tel.: +86-188-5110-9232

Received: 22 June 2019; Accepted: 30 July 2019; Published: 1 August 2019

Abstract: The widespread penetration of hydrogen in mainstream energy systems requires hydrogen production processes to be economically competent and environmentally efficient. Hydrogen, if produced efficiently, can play a pivotal role in decarbonizing the global energy systems. Therefore, this study develops a framework which evaluates hydrogen production processes and quantifies deficiencies for improvement. The framework integrates slack-based data envelopment analysis (DEA), with fuzzy analytical hierarchy process (FAHP) and fuzzy technique for order of preference by similarity to ideal solution (FTOPSIS). The proposed framework is applied to prioritize the most efficient and sustainable hydrogen production in Pakistan. Eleven hydrogen production alternatives were analyzed under five criteria, including capital cost, feedstock cost, O&M cost, hydrogen production, and CO_2 emission. FAHP obtained the initial weights of criteria while FTOPSIS determined the ultimate weights of criteria for each alternative. Finally, slack-based DEA computed the efficiency of alternatives. Among the 11, three alternatives (wind electrolysis, PV electrolysis, and biomass gasification) were found to be fully efficient and therefore can be considered as sustainable options for hydrogen production in Pakistan. The rest of the eight alternatives achieved poor efficiency scores and thus are not recommended.

Keywords: hydrogen production processes; economic viability; environmental efficiency; sustainable energy; multi-criteria analysis

1. Introduction

Hydrogen is identified as the most critical and indispensable energy alternative that forms a viable option for the decarbonization of the global energy system [1]. A growing body of literature suggests five essential factors that enable hydrogen to become a future low-carbon energy pathway [2,3]. Firstly, hydrogen is the universe's most abundant element [4]. Secondly, hydrogen has a massive potential to reduce greenhouse gases (GHGs) [5]. Thirdly, it is a versatile energy carrier that can operate across various sectors, including industry [6], transport [7], heat [8], and electricity [9]. Fourthly, it can offset electricity as zero-carbon energy that can be easily transported and stored [10]. Lastly, it enhances energy security by reducing dependence on fossil fuel [11]. The objective of this study is to provide a framework to assess the feasibility of hydrogen production processes for mass exploitation of this abundant natural resource.

A variety of energy sources and processes can be used to produce hydrogen. Currently, 96% of hydrogen is produced from fossil fuels using steam methane reforming (SMR) process. Three major fossil fuels used for hydrogen production are natural gas (48%), oil (30%), and coal (18%) [3]. Hydrogen production comprises extraction and isolation of hydrogen in the shape of independent molecules, at a purity level that is necessary for a given application. The methods of hydrogen production typically rely on starting point, and the presently leading technique of production from methane can only be reasonable if the energy is firstly contained in methane or can be easily transformed to methane. Therefore, in the case of fossil fuels, the hydrogen production from natural gas is relatively easy, from oil is a little bit more intricate, while from coal needs initially high-temperature gasification [12].

For hydrogen production from electricity, the process of electrolysis is commonly used. Currently, this process produces the rest of the 4% of total hydrogen [3]. The electrolysis process which uses renewable electricity is called renewable electrolysis. The two most common renewable electrolysis methods are wind electrolysis and solar electrolysis. Renewable electrolysis offers some additional and promising benefits such as hydrogen fuels storage that can reinforce increased penetration of renewable energy. Other renewable energy sources (RES) such as biomass is also used for hydrogen production. However, unlike other RES, biomass requires some special treatment, depending on the form of biomass feedstock. For instance, at high temperature, direct decomposition of water or photo-induced are considered, while at low temperature, more complicated and multistep processes are required, such as the ones offered by steam from concentrating solar power plants or nuclear reactors [12].

The wide-ranging availability of hydrogen production processes complicates the decision-making regarding the selection of the most sustainable process [13]. These processes use massive inputs, such as capital cost, feedstock cost, and operation and maintenance (O&M) cost, to produce hydrogen, while simultaneously producing undesirable outputs, such as GHGs emissions, as a byproduct. Therefore, to enable decision makers to choose the best hydrogen production process, it is crucial to evaluate the economic viability and environmental efficiency of various hydrogen production processes [5]. This task can be achieved by using the environmental data envelopment analysis (DEA), which is the most common method of efficiency evaluation when undesirable outputs are involved. However, the DEA model calculates the preference weights of variables (i.e., inputs, desirable outputs, and undesirable outputs) automatically, while ignoring the relative importance of these variables to each other in the calculation [14]. Hence, DEA considers the equal importance of each variable. However, this is contrary to reality, in which the preference of variables changes depending on stakeholders' considerations.

Therefore, to address the problem of equal weights, this study develops a framework which applies multi-criteria decision analysis (MCDA) techniques to determine the importance of each variable, before assessing the economic and environmental efficiency of hydrogen production processes. Two most popular MCDA techniques, i.e., fuzzy analytical hierarchy process (FAHP), and the fuzzy technique for order of preference by similarity to ideal solution (FTOPSIS) were combined with slack-based environmental DEA to accomplish the task. Conventional analytical hierarchy process (AHP) and technique for order of preference by similarity to ideal solution (TOPSIS) could also have been used. However, the techniques lack in dealing with the vagueness and bias involved in stakeholders' considerations. Fuzzy, on the other hand, is proficient in handling the uncertainty and vagueness involved in the experts' feedback [15]. Likewise, using slack-based environmental DEA, instead of a simple environmental DEA, is more useful as it provides the information of slack-variables (i.e., excess of inputs and undesirable outputs, and the shortfall of desirable outputs) and overcoming slacks can help to improve efficiency performance [16]. The proposed framework is employed to analyze the case of Pakistan, which is an energy deficient and environmentally vulnerable country.

Initially, the hydrogen production processes also referred to as alternatives, available in Pakistan, were shortlisted. After that, variables also termed as criteria, used to evaluate the performance of those processes/alternatives were finalized. Three input criteria (capital cost, O&M cost, and feedstock cost), one desirable output criteria (hydrogen production), and one undesirable output criteria (CO_2 emission) were selected for the analysis. FAHP was used to determine the initial weights of each criterion. After obtaining the initial weights, FTOPSIS was employed to determine the ultimate weights of criteria for each alternative. Finally, slack-based environmental DEA was used to compute efficiency scores of alternatives and rank them according to their scores.

The rest of the study proceeds as follows: Section 2 provides the literature review of MCDA techniques used in the decision-making related to hydrogen production processes. Section 3 delineates the proposed methodology. Section 4 applies the proposed methodology to prioritize the most sustainable hydrogen production process in Pakistan. Section 5 presents and discusses the results of the study. The final section concludes the study.

2. MCDA for Hydrogen Selection

Since the assessment of hydrogen production processes and the decision-making related to the selection of the most viable processes involve multi-dimensional criteria, MCDA techniques have been widely used in the relevant literature. Acar et al. [17] used fuzzy hesitant AHP to conduct the sustainability analysis of various hydrogen production methods including grid electrolysis, photovoltaic (PV) electrolysis, wind electrolysis, solar thermochemical water splitting, nuclear thermochemical water splitting, and photo-electrochemical cells. The selected methods were evaluated based on five criteria, i.e., technical performance, economic performance, environmental performance, social performance, and reliability. The results of the study showed that grid electrolysis is the most sustainable hydrogen production option.

Ren and Toniolo [18] proposed a novel MCDA method to rank sustainable hydrogen production pathways by combining interval evaluation based on distance from average solution (EDAS) and improved decision making trial and evaluation laboratory (DEMATEL). They studied four hydrogen production pathways, including SMR, coal gasification, biomass gasification, and wind electrolysis. Biomass gasification was found to be the most sustainable pathway followed respectively by SMR, wind electrolysis, and coal gasification.

Ren et al. [19] applied FAHP and FTOPSIS to prioritize the role of various hydrogen production technologies for developing a hydrogen economy in China. Hydrogen production technologies assessed in the study include SMR, coal gasification with CO_2 capture and storage, nuclear-based high-temperature electrolysis, biomass gasification, and hydropower electrolysis. The selected technologies were assessed based on 10 criteria under four aspects, including technical, economic, environmental, and socio-political. The results showed hydropower-based water electrolysis and coal gasification with CO_2 capture and storage as the two most important hydrogen production technologies, among others for establishing a hydrogen economy in China.

Yu [20] developed a decision-making model for the selection of hydrogen production technologies in China. The model was established based on interval-valued intuitionist fuzzy set theory. The study assessed three hydrogen production technologies, including coal gasification, water electrolysis using hydropower, and nuclear based high-temperature electrolysis. The evaluation used three criteria, inducing the degree of political support, economic performance, and social performance. Nuclear based high-temperature electrolysis was considered as the best technology for hydrogen production while the remaining two technologies were found to be least satisfactory.

Ren et al. [21] proposed a novel fuzzy multi-actor MCDA model, which enabled multiple decision-making groups to use linguistic variables to assess the sustainability of four biomass-based hydrogen production processes including biomass pyrolysis, biomass fermentative hydrogen production, biomass gasification, and biomass supercritical water gasification. The authors used 15 criteria related to economic, technological, socio-political, and environmental aspects of the assessment. The results of the study reported biomass gasification as the most sustainable process and fermentative hydrogen production as the least sustainable option.

Pilavachi et al. [22] used AHP to prioritize seven hydrogen production technologies, including SMR, coal gasification, biomass gasification, partial oxidation of hydrocarbons, wind electrolysis, PV electrolysis, and hydropower electrolysis. The prioritization was done based on five criteria, including

CO_2 emissions, capital cost, operation and maintenance, hydrogen production cost, and feedstock cost. The assessment ranked PV electrolysis, wind electrolysis, and hydropower electrolysis higher than the conventional technologies, SMR, coal gasification, partial oxidation of hydrocarbons, and biomass gasification.

The above literature provides enough evidence regarding the extensive utilization of MCDA techniques for the assessment of hydrogen technologies. The above-reviewed studies take into account various aspects and criteria to prioritize different methods of hydrogen production. One crucial point that is missing in the past studies is the lack of proper treatment of undesirable outputs, which are byproducts in the hydrogen production processes and can influence the environmental efficiency of these processes. To tackle undesirable outputs in the analysis, this study integrates MCDA with environmental DEA to rank hydrogen production processes.

3. Methodology

The proposed methodology combines FAHP, FTOPSIS, and slack based environmental DEA to develop a framework for the selection of relatively efficient hydrogen production technologies. Figure 1 presents the flowchart of the research design. The steps involved in the proposed approach are:

i. Finalize hydrogen production technologies (alternatives) to be evaluated.
ii. Select variables (criteria) and categorize them into inputs, desirable outputs, and undesirable outputs.
iii. Employ FAHP to compute initial weights of criteria.
iv. Use FAHP weights to compute the final of criteria for each alternative using FTOPSIS.
v. Use final weights in slack-based environmental DEA to obtain the final ranking of alternatives.

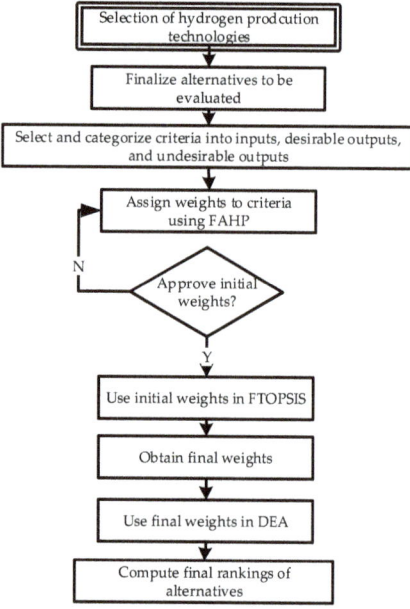

Figure 1. Flowchart of research design.

3.1. FAHP

Saaty introduced AHP as a quantitative method of multi-criteria decision analysis [23]. The Saaty AHP has some limitations because it can only be applied where there is no uncertainty, the environment is crisp, the selection of judgement is subjective, and the judgmental scale is unbalanced. Therefore, Fuzzy approach is integrated with AHP to extend the latter's applicability. The FAHP proficiently deals with imprecise and uncertain judgment of experts on the field by using linguistic variables [24]. Definition of fuzzy operations is as follows:

If $\widetilde{D_1} = (b1_1, b2_1, b3_1)$ and $\widetilde{D_2} = (b1_2, b2_2, b3_2)$ are representing two triangular fuzzy numbers (TFNs) then algebraic operations can be expressed as follows [25]

$$\widetilde{D_1} \oplus \widetilde{D_2} = (b1_1 + b1_2, b2_1 + b2_2, b3_1 + b3_2), \quad (1)$$

$$\widetilde{D_1} \odot \widetilde{D_2} = (b1_1 - b3_2, b2_1 - b2_2, b3_1 - b1_2), \quad (2)$$

$$\widetilde{D_1} \otimes \widetilde{D_2} = (b1_1 b1_2, b2_1 b2_2, b3_1 b3_2), \quad (3)$$

$$\widetilde{D_1} \oslash \widetilde{D_2} = \left(\frac{b1_1}{b3_2}, \frac{b2_1}{b2_2}, \frac{b3_1}{b1_2}\right), \quad (4)$$

$$a \otimes \widetilde{D_1} = (ab1_1, ab2_1, ab3_1) \text{ where } a > 0, \quad (5)$$

$$\widetilde{D_1}^{-1} = \left(\frac{1}{b3_1}, \frac{1}{b2_1}, \frac{1}{b1_1}\right). \quad (6)$$

The FAHP is applied according to the method proposed by [26] as follows:

$$T_{g_i}^1, T_{g_i}^2, T_{g_i}^3, \ldots, T_{g_i}^m, \quad (7)$$

where $T_{g_i}^j$ is ($j = 1, 2, 3, 4, 5, 6, \ldots, m$) TFNs provided in Table 1, and g_i is the goal set ($i = 1, 2, 3, 4, 5, 6, \ldots, n$).

Table 1. Triangular fuzzy numbers linguistic scale.

TFN	Linguistic Variable	TFN Scale
1	Equally important	(1, 1, 1)
2	Weakly advantage	(1, 2, 3)
3	Not a bad advantage	(2, 3, 4)
4	Preferred	(3, 4, 5)
5	Good advantage	(4, 5, 6)
6	Fairly good advantage	(5, 6, 7)
7	Very good advantage	(6, 7, 8)
8	Absolute advantage	(7, 8, 9)
9	Perfect advantage	(8, 9, 10)

FAHP involves the following steps:

Step 1: Use TFNs to construct pairwise comparison matrixes of attributes.
Step 2: Fuzzy synthetic extent (S_i) value of ith element can be defined as:

$$S_i = \sum_{j=1}^{m} T_{g_i}^j \times \left[\sum_{i=1}^{n} \sum_{j=1}^{m} T_{g_i}^j\right]^{-1}$$

$$\sum_{j=1}^{m} T_{g_i}^j = \left(\sum_{j=1}^{m} b1_{ij}, \sum_{j=1}^{m} b2_{ij}, \sum_{j=1}^{m} b3_{ij}\right) \quad (8)$$

$$\left[\sum_{i=1}^{n} \sum_{j=1}^{m} T_{g_i}^j\right]^{-1} = \left(\frac{1}{\sum_{n}^{i=1} \sum_{m}^{j=1} b3_{ij}}, \frac{1}{\sum_{n}^{i=1} \sum_{m}^{j=1} b2_{ij}}, \frac{1}{\sum_{n}^{i=1} \sum_{m}^{j=1} b1_{ij}}\right)$$

Step 3: Comparison of the obtained values of S_i, and compute the possibility degree $S_j = (b1_j, b2_j, b3_j) \geq S_i = (b1_i, b2_i, b3_i)$. Following is the equivalent expression:

$$V(S_j \geq S_i) = u_{S_j}(d) = \begin{cases} 1, & \text{in case of } b2_j \geq b2_i \\ 0, & \text{in case of } b1_i \geq b3_j \\ \frac{b1_i - b3_j}{(b2_j - b3_j) - (b2_i - b1_i)}, & \text{otherwise} \end{cases}, \quad (9)$$

where d denotes highest ordinate point between u_{S_j} and u_{S_i}. Both $V(S_j \geq S_i)$ and $V(S_i \geq S_j)$ values are required to compare S_j and S_i.

Step 4: Computation of the minimum possibility degree $d(i)$ of $V(S_j \geq S_i)$ for $(ij = 1,2,3,4,5,\ldots,k)$ can be defined as

$$\begin{array}{c} V(S \geq S_1, S_2, S_3, S_4, S_5 \ldots, S_k), \\ \text{for } (i = 1,2,3,4,5 \ldots, k) \\ = V[(S \geq S_1), (S \geq S_2), \text{and} \ldots (S \geq S_k)] = \min V(S \geq S_i) \\ \text{for } (i = 1,2,3,4,5, \ldots, k) \end{array} \quad (10)$$

If we assume that

$$d'(A_i) = \min V(S \geq S_i); \text{ for } (i = 1,2,3,4,5 \ldots, k). \quad (11)$$

Then, we can define the weight vector by

$$W' = (d'(A_1), d'(A_2), d'(A_3), d'(A_4), \ldots, d'(A_n))^T, \quad (12)$$

where A_1 for $(i = 1,2,3,4,5 \ldots, n)$ are the n objects

Step 5: Normalize weight vectors as below

$$W = (d(A_1), d(A_2), d(A_3), d(A_4), \ldots, d(A_n))^T. \quad (13)$$

W denotes a non-fuzzy number.

3.2. FTOPSIS

TOPSIS is another widely used MCDM technique to solve decision-making problems in a variety of fields. TOPSIS is a linear weighting method proposed by [27]. The method was proposed initially in its crisp version. TOPSIS chooses an alternative that has the longest distance from a negative ideal solution and the shortest distance from a positive ideal solution. This method describes an index that measures the similarity to the positive ideal solution and differences to the negative ideal solution. Finally, the method selects an alternative which has more similarity to the positive ideal solution [28]. The classical TOPSIS approach uses crisp values to assign individual preferences. However, in reality, it often becomes hard for decision-makers to assign a precise performance score. Therefore, a better technique is considering vagueness and uncertainty instead of crisp values. Fuzzy integrates uncertainty in decision making, therefore, the FTOPSIS method is more appropriate to obtain solutions of real-life problems [29]. The FTOPSIS, in this paper, is used in the following steps:

Step 1: Assign a rating to linguistic variables in relation to criteria and construct fuzzy matrixes for alternatives. Table 2 lists scoring used to rate linguistic variables.

Step 2: Construct fuzzy decision/performance matrix

$$\widetilde{D} = \begin{matrix} A_1 \\ \vdots \\ A_m \end{matrix} \begin{bmatrix} C_1 & \cdots & C_n \\ \widetilde{x}_{11} & \cdots & \widetilde{x}_{1n} \\ \vdots & \ddots & \vdots \\ \widetilde{x}_{m1} & \cdots & \widetilde{x}_{mn} \end{bmatrix} \quad (14)$$

$$i = 1, 2, 3, 4, 5, \ldots, m; \ j = 1, 2, 3, 4, 5, \ldots, n$$
$$\widetilde{x}_{ij} = \tfrac{1}{K}\left(\widetilde{x}_{ij}^1 \oplus \cdots \oplus \widetilde{x}_{ij}^k \oplus \cdots \oplus \widetilde{x}_{ij}^K\right),$$

where $\widetilde{x}_{ij}^k$ denotes performance rating of A_i alternative with respect to C_j criteria evaluated by kth decision matrix, and $\widetilde{x}_{ij}^k = \left(b1_{ij}^k, b2_{ij}^k, b3_{ij}^k\right)$.

Step 3: Compute the normalized fuzzy decision/performance matrix. Data is normalized to obtain a comparable measure by using linear scale transformation as below

$$\widetilde{B} = \left[\widetilde{p}_{ij}\right]_{m \times n}, \quad (15)$$

where ($i = 1,2,3,4,5, \ldots, m$) and ($j = 1,2,3,4,5, \ldots, n$),

$$\widetilde{p}_{ij} = \left(\frac{b1_{ij}}{b3_j^*}, \frac{b2_{ij}}{b3_j^*}, \frac{b3_{ij}}{b3_j^*}\right) \text{ and } b3_j^* = \max b3_{ij} \text{ (benefit criteria)},$$

$$\widetilde{p}_{ij} = \left(\frac{b1_j^-}{b3_{ij}}, \frac{b1_j^-}{b2_{ij}}, \frac{b1_j^-}{b1_{ij}}\right) \text{ and } b1_j^- = \min b1_{ij} \text{ (cost criteria)}.$$

Step 4: Compute the weighted normalized matrix using the given equations:

$$\widetilde{V} = \left[\widetilde{v}_{ij}\right]_{m \times n} \text{ where } i = 1, 2, 3, 4, 5, \ldots, m \text{ and } j = 1, 2, 3, 4, 5, \ldots, n, \quad (16)$$

$$\widetilde{v} = \widetilde{p}_{ij} \otimes w_{ij}, \quad (17)$$

where w_{ij} shows the weight of C_j criterion. Criteria weights used here are obtained from the FAHP method.

Step 5: Find the fuzzy positive ideal solution (FPIS) and fuzzy negative ideal solution (FNIS) respectively as follows

$$A^+ = \left\{v_1^+, \ldots, v_n^+\right\}, \text{ where } v_j^+ = \left\{\max(v_{ij}) \text{ in case of } j \in J; \min(v_{ij}) \text{if } j \in J'\right\}, j = 1, 2, 3, 4, 5, \ldots, n, \quad (18)$$

$$A^- = \left\{v_1^-, \ldots, v_n^-\right\}, \text{ where } v_j^- = \left\{\min(v_{ij}) \text{ in case of } j \in J; \max(v_{ij}) \text{if } j \in J'\right\}, j = 1, 2, 3, 4, 5, \ldots, n. \quad (19)$$

Step 6: Determine the distance of alternatives from FPIS and FNIS as follows

$$\widetilde{d}_i^+ = \left\{\sum_{j=1}^n \left(v_{ij} - v_{ij}^+\right)^2\right\}^{0.5}, \text{ for } i = 1, 2, 3, 4, 5, \ldots, m$$
$$\widetilde{d}_i^- = \left\{\sum_{j=1}^n \left(v_{ij} - v_{ij}^-\right)^2\right\}^{0.5}, \text{ for } i = 1, 2, 3, 4, 5, \ldots, m \quad (20)$$

Step 7: Compute closeness coefficient (CC_i) as follows

$$CC_i = \frac{d_i^-}{d_i^+ + d_i^-}, \text{ for } i = 1,2,3,4,5,\ldots,m; \text{ and } C_i \in (0,1). \tag{21}$$

Step 8: Obtain final ranking of alternatives using CC_i values.

Table 2. Ratings of linguistic variables.

Linguistic Variables	TFNs
Moderate low (ML)	(0, 0, 1)
Low (L)	(0, 1, 3)
Slightly lower (SL)	(1, 3, 5)
Fair (F)	(3, 5, 7)
Slightly higher (SH)	(5, 7, 9)
High (H)	(7, 9, 10)
Moderate High (MH)	(9, 10, 10)

3.3. Environmental Data Envelopment Analysis (DEA)

Data envelopment analysis (DEA) is a nonparametric measure of efficiency. DEA does not need any profound information of production process of "decision-making unit" (DMU) [30]. For DEA efficiency evaluation, it is suffice to select proper inputs, outputs, and undertake some assumptions regarding the technological structure pertaining to disposability, convexity, and returns to scale [31].

The standard DEA, as described in [32], depends on the assumption that inputs need to be minimized and outputs need to be maximized. However, as mentioned in the seminal work of Koopman [33], the production process can also produce undesirable outputs (e.g., wastes or pollutants) as byproducts from an environmental perspective. The classical DEA models do not take into account asymmetry between desirable and undesirable outputs and therefore result in erroneous calculations and biased performance assessment.

Since the hydrogen production process also produces undesirable outputs (e.g., CO_2 emission), we rely on directional measures [34] to incorporate undesirable outputs in classical DEA efficiency models. Directional measures treat both desirable and undesirable outputs differently.

Incorporation of desirable and undesirable output requires a redefinition of the production function. For instate, the initial vector of $i = 1,2,3,4,\ldots,s$ outputs $y \in \mathbb{R}_{++}^s$ redefined into $y = y^d + y^u$, with $y^d \in \mathbb{R}_{++}^q$ desirable outputs and $y^u \in \mathbb{R}_{++}^r$ as undesirable. Therefore the corresponding reference technology $P_{CRS} = \{(x, y^d, y^u) | x \geq X\lambda, y^d \leq Y\lambda, y^u \leq Y\lambda, \lambda \geq 0\}$, shows weak disposability of undesirable outputs (for more details see [35]). In such a case, the observation of directional efficiency measure (x, y^d, y^u) along a pre-assigned direction corresponding to the vector of output $g_y = y^d y^u \neq 0_{m+s}$, corresponds to the solution of the following model:

$$\max \beta. \tag{22}$$

Subject to

$$\begin{aligned} X\lambda &\leq x \\ Y^d \lambda &\geq y^d + \beta y^d \\ Y^u \lambda &\geq y^u + \beta y^u \\ \max\{y_i^u\} &\geq y^u - \beta y^u \\ \lambda &\geq 0 \end{aligned} \tag{23}$$

Here the optimal solution corresponds to β^*_{CRS}, the observation is directional efficient if $\beta^*_{CRS} = 0$, with $\lambda = 1, \lambda_j = 0 \ (j \neq 0)$. Otherwise, $\beta^*_{CRS} > 0$ shows inefficiency and (x, y^d, y^u) outdoes (x, y^d, y^u).

The model also calculates non-directional slacks, checking for excess in inputs and undesirable outputs or any shortfall in desirable outputs.

4. Case Study

The developed framework was applied to solve the decision-making problem of selecting hydrogen production technologies in Pakistan. It is very significant to consult professional and experienced experts while applying any MCDM approach. Initially, 20 experts were asked to participate in the study. The questionnaire survey was distributed to experts through webmail service. However, four experts could not participate, so the final list included 16 experts. Each of the expert were asked to weight the importance using pairwise comparison matrix of the criteria, sub- criteria and alternatives. The experts who participated belong to academia, energy department, economists, stakeholders, and research specialists. The demographic information of the experts is presented in Table 3.

Table 3. The demographic information of experts.

Number of Experts	Designation	Organization
2	Associate Professor	MUET, Jamshoro, Pakistan
2	Assistant Professor	UoS, Jamshoro, Pakistan
1	Assistant Director	HESCO, Hyderabad
1	Manager	NTDC, Islamabad
1	Deputy Director	MoPW, Islamabad
3	Consultants	Green consultancy
4	PhD Scholars	Nanjing University of Aeronautics and Astronautics
1	Assistant Director	PCRET, Islamabad
1	Energy specialist	PAEC, Lahore

The reason Pakistan was selected as a case study is that Pakistan is an energy deficient country that is struggling to adopt new approaches to address its decades-long energy crisis [36]. Being new to the concept of the hydrogen economy, Pakistan should not take the same inefficient steps, which developed countries used during their initial phase of transition to a hydrogen economy. Instead, Pakistan, following the lessons learnt by developed countries, should leapfrog the inefficient steps and adopt efficient technologies and processes for implementing a hydrogen economy. The case study is implemented according to the proposed framework as follows:

4.1. Selection of Alternatives

Hydrogen is an abundantly available element on the earth. However, it is almost always found as a component of other compounds. For instance, hydrogen is found in water (H_2O), and if hydrogen is to be used as fuel, then it must be separated from oxygen [37]. Apart from water, other diverse sources that can produce hydrogen include fossil fuels, biomass, and several other domestic sources. Energy efficiency, environmental impacts, and cost of hydrogen depend on the process through which it is produced [38].

There are numerous ways to produce hydrogen. However, this study shall only consider technological processes which can be applied in the Pakistani context. These processes include thermochemical, electrolysis, direct solar water spiriting, and biological process [39]. These technologies have great scope in Pakistan after their recent breakthrough. However, the transformation from fossil fuel economy to hydrogen needs solutions of various complex technological challenges. The provision of cost-competitive hydrogen energy of adequate quality and quantity is the basis of hydrogen economy. Therefore, we analyzed available hydrogen production processes to find the best hydrogen production process in terms of environmentally clean and economically viable. Figure 2 shows 11 selected alternatives under each process. These alternatives are also briefly explained as follows.

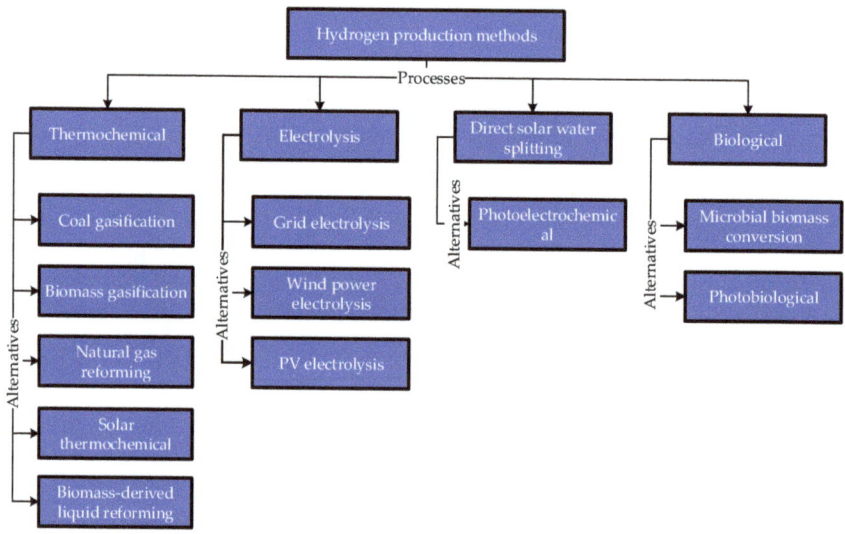

Figure 2. Hydrogen production alternatives.

4.1.1. Thermochemical Process

Some thermochemical processes use chemical reactions and energy to release hydrogen from the molecular structure of various organic materials such as coal, biomass, and natural gas. Other processes produce hydrogen from feedstock by combining heat with closed-chemical cycles. The most common and viable thermochemical processes are coal gasification, biomass gasification, and natural gas reforming [13].

Coal Gasification

Coal is a highly variable and complex substance that can be transformed into a range of products. Coal gasification is a method that converts coal into chemicals, liquid fuels, and hydrogen. Coal is initially reacted with steam and oxygen under extreme temperature and pressure to create syngas, a mixture of hydrogen (H_2) and carbon monoxide (CO) [40]. Once impurities are separated from the synthesis gas, water-gas reaction reacts CO to produce additional hydrogen and CO_2. Later, the separation system removes hydrogen, and subsequently captures and stores the highly concentrated carbon stream.

Biomass Gasification

Biomass, a renewable resource, includes animal dung, agriculture crop residue, forest residue, crops grown for energy use (e.g., willow trees or switchgrass), and organic municipal solid waste. This renewable resource can produce hydrogen and other byproducts through gasification. The process of biomass gasification uses a controlled amount of oxygen, heat, and steam to convert biomass into hydrogen, CO, and CO_2, without combustion. The CO then reacts with water to produce CO_2 and additional hydrogen through a water-gas shift reaction. Absorbers or special membranes are used to split hydrogen from the gas stream [41].

Natural Gas Reforming

Natural gas reforming develops upon the existing infrastructure of gas delivery. It is an advanced and mature hydrogen production process. Methane (CH_4) in natural gas can produce hydrogen through the thermal process. The primary thermal ways to convert CH_4 into hydrogen involve reaction with either oxygen (partial oxidation), steam (steam reforming), or a sequence of both (autothermal reforming) [13]. Practically, gas mixtures containing CO, CO_2, and CH_4 require further processing. The reaction of carbon monoxide with steam over a catalyst generates an extra amount of hydrogen and carbon dioxide, and only after purification, high-purity hydrogen is obtained. Most often, CO_2 vents into the atmosphere; however, numerous options exist to capture it for sequestration.

Solar Thermochemical

Thermochemical water splitting drives a series of chemical reactions under extreme temperature (500–2000 °C) that split water into hydrogen and oxygen. Chemicals used in this process are recycled within each chemical reaction creating closed loops that only use water to generate oxygen and hydrogen. This process produces low or no greenhouse gases and therefore is considered as a long-term technology pathway [42]. Numerous cycles of solar thermochemical water splitting have been examined for hydrogen production, each having different operation conditions, challenges, and production opportunities. In fact, the literature shows more than 300 cycles of solar thermochemical water splitting [43]. The two most common cycles include the direct (two-stem cerium oxide) and the hybrid (copper chloride cycle). Figure 2 illustrates the schematic of these cycles. Direct cycles have fewer steps and are typically less complicated; however, they require a higher temperature compared with complicated hybrid cycles.

Biomass-Derived Liquid Reforming

The liquid obtained from biomass includes bio-oils, cellulosic ethanol, and other liquid biofuels can be reformed for hydrogen production [44]. These liquids are easier to transport compared to biomass feedstock allowing hydrogen reforming at fueling stations, stationary power cites, or semi-central production facilities. The large and centralized liquid producing facilities can be established near the biomass source to exploit economies of scale and reduce transportation cost of solid biomass feedstock [45].

The process of hydrogen reforming from biomass-derived liquids is similar to natural gas reforming, and involves the following three steps:

i. In the presence of a catalyst, liquids are reacted with steam at high heat to form reformate gas composed mainly of CO, CO_2, and hydrogen.
ii. Excess amount of hydrogen and CO_2 are produced by reacting CO with high-heat steam in the "water-gas shift reaction."
iii. In the final step, hydrogen is parted and purified.

4.1.2. Electrolysis Process

Electrolysis uses electricity to split water into oxygen and hydrogen. Electrolysis is of interest as a promising source because it uses water to produce hydrogen, and water is abundantly available as compared to hydrocarbons. The reaction of splitting water takes place in the electrolyzer. The sizes of electrolyzer vary. Small sized electrolyzers are appropriate for small scale hydrogen production. Large sized are well-suited for centralized production facilities that could be connected directly to any form of electricity (renewable or non-renewable) production [46]. Electrolyzers, like fuel cells, have an anode and a cathode detached by an electrolyte. Functions of different electrolyzers are slightly different from each other, mainly because of being built up of different kinds of electrolyte material. Electrolysis technology is well-developed and commercially available [39].

Grid Electrolysis

The grid electrolysis uses conventional electricity to produce hydrogen. In this process, electrolysis is connected to the electricity grid. This process is a fast and cheap way of transitioning to a hydrogen economy [47]. Currently, grid electricity costs Rs. 20.79 kWh [36]. However, this option is not viable in remote areas with lack of access to reliable electricity. In addition, even though the process of electrolysis is itself clean, however, the process of grid electrolysis contributes greenhouse gas emissions because most of the grid electricity in Pakistan is produced from fossil fuels [48].

Wind Electrolysis

The process of wind electrolysis is quite similar to the grid electrolysis except for one difference: the electrolyzer in wind electrolysis is connected to the electricity produced using wind turbines. Wind electrolysis is a viable technique to produce clean hydrogen. The process enables the better use of indigenous renewable energy sources. Wind electrolysis, due to being a green method, can help to reduce greenhouse gas emissions while integrating a larger share of clean energy into the electric grid [48]. For a larger penetration of renewable energy, wind electrolysis for hydrogen production must be cost competitive. Besides low production cost, transportation and storage costs factors should also be taken into the final hydrogen production cost. These factors necessitate the investigation of wind class sites, considering the geographical distance from the end-user [49]. Currently, Europe is the leader in the field of hydrogen production via wind electrolysis. The European Union has recently implemented a successful demonstration of wind electrolysis based hydrogen production project in Spain and Greece. The project involved hydrogen storage, desalination technology, and fuel cells, and provided renewable hydrogen energy for power supply, energy storage, and supply of fresh water [50].

PV Electrolysis

The large solar energy resource potential, the advancement in its technology, and the rapidly falling cost drive the rapid growth of utility-scale solar electricity generating plants [51]. The maturity in solar electricity generation provides a viable opportunity for hydrogen generation from solar electrolysis. Solar electrolysis is the process of producing hydrogen via solar splitting water. The solar electrolysis presents a promising solution to the challenges of hydrogen storage, transportation, and generation without producing harmful byproducts [52].

4.1.3. Direct Solar Water Splitting Process

The process of direct solar water splitting produces hydrogen by splitting water with the help of light energy. Currently, this process is at the early stage of research. However, it provides great potential and a long-term sustainable option for hydrogen production with minimum impacts on the environment [13]. Below is the process of solar water splitting:

Photoelectrochemical

Photoelectrochemical water splitting produces hydrogen from splitting water using specialized semiconductors and sunlight. These specialized semiconductors are called photoelectrochemical materials, which use energy from sunlight to directly separate water molecules into oxygen and hydrogen. The process of photoelectrochemical is a long-term hydrogen production pathway with the lowest greenhouse gas emissions [53].

4.1.4. Biological Process

Microbes such as microalgae and bacteria can produce hydrogen via biological reactions by using organic material and sunlight. Biological processes, similar to direct solar water splitting processes, are also at an early stage of research. Biological processes of hydrogen production provide a sustainable and low-carbon option for hydrogen production [54]. Given below are two common biological processes. These processes are found to be less energy intensive and more environmental friendly as compared to electrochemical and thermochemical processes [55].

Microbial Biomass Conversion

The microbial process uses the ability of the microorganism to consume and digest biomass and produce hydrogen. Microbial systems can be suitable for central, semi-central, or distributed hydrogen production depending on the feedstock used [56]. There are different ways of microbial process. The fermentation-based process uses microorganisms, such as bacteria, to convert organic matter into hydrogen. The organic matter can be raw biomass sources, refined sugar, and even wastewater. This method is sometimes called the dark fermentation method due to no requirement of light in the process. The direct hydrogen fermentation process uses microbes themselves to produce hydrogen [57]. Microbes break complex molecules via various pathways. These pathways generate byproducts, and the enzymes combine these byproducts to produce hydrogen. Researchers are paying adequate attention to improve the yield (using the same amount of organic matter) and the speed of hydrogen production from fermentation [58]. In fact, the yield has been improved. There used to be a ceiling for hydrogen production (4 mol H_2/mol glucose). Recently, strains have been developed showing hydrogen production can be up to 8 mol H_2/mol glucose [59].

Photobiological

In the photobiological process, microorganisms, such as cyanobacteria and microalgae, use sunlight to convert water, and sometimes organic matter, into oxygen and hydrogen ions [60]. The hydrogen ions, once combined via direct and indirect ways, are released as hydrogen gas. Some photosynthetic microbes use sunlight to disintegrate organic matter to produce hydrogen. This process is called the photo-fermentative process of hydrogen production. Recently, the photobiological process has significantly progressed, and is being considered as a mature technology. Few challenges that make this process unviable at this time include low rates of hydrogen production and solar to hydrogen efficiency [50].

4.2. Compute Initial Weights of Criteria Using FAHP

Six variables were selected for the analysis. These variables include three inputs (capital cost, operation and maintenance O&M cost, and feedstock cost), one desirable output (amount of hydrogen production in kg), and one undesirable output (CO_2 emission).

The first step was to compute initial weights using FAHP includes incorporating experts' judgments into the pairwise matrix, which is given in Table 4.

Later, the fuzzy synthetic (S_i) values of variables were calculated using Equation (8) as below:

S_1(Capital cost) =
$(4.972, 6.767, 8.975) \otimes (0.029, 0.038, 0.049)$
$= (4.972 * 0.029 * 6.767 * 0.038 * 8.975 * 0.049)$
$= (0.143, 0.256, 0.44)$
S_2(CO_2 emission) =
$(2.948, 3.69, 4.799) \otimes (0.029, 0.038, 0.049)$
$= (2.948 * 0.029 * 3.69 * 0.038 * 4.799 * 0.049)$
$= (0.085, 0.139, 0.235)$
S_3(Feedstock cost) =
$(4.203, 5.498, 7.301) \otimes (0.029, 0.038, 0.049)$
$= (4.203 * 0.029 * 5.498 * 0.038 * 7.301 * 0.049)$
$= (0.121, 0.208, 0.358)$
S_4(O&M) =
$(3.392, 4.191, 5.353) \otimes (0.029, 0.038, 0.049)$
$= (3.392 * 0.029 * 4.191 * 0.038 * 5.353 * 0.049)$
$= (0.098, 0.158, 0.263)$
S_5(Hydrogen production) =
$(4.87, 6.319, 8.262) \otimes (0.029, 0.038, 0.049)$
$= (4.87 * 0.029 * 6.319 * 0.038 * 8.262 * 0.049)$
$= (0.14, 0.239, 0.405)$

These S_i values were compared to calculate the possibility degree $S_j = (b1_j, b2_j, b3_j) \geq S_i = (b1_i, b2_i, b3_i)$. The comparison of $S_j = (b1_j, b2_j, b3_j) \geq S_i = (b1_i, b2_i, b3_i)$ is presented in Table 5.

Table 4. Pairwise matrix of the fuzzy analytical hierarchy process (FAHP).

	Capital Cost	CO₂ Emission	Feedstock Cost	O&M Cost	Hydrogen Production
Capital Cost	1, 1, 1	1.19, 1.75, 2.43	0.82, 1.14, 1.53	1.19, 1.76, 2.44	0.77, 1.12, 1.58
CO₂ emission	0.41, 0.57, 0.84	1, 1, 1	0.48, 0.7, 1.04	0.63, 0.84, 1.11	0.42, 0.58, 0.81
Feedstock Cost	0.65, 0.88, 1.22	0.96, 1.42, 2.07	1, 1, 1	0.86, 1.18, 1.6	0.73, 1.01, 1.41
O&M Cost	0.41, 0.57, 0.84	0.9, 1.19, 1.58	0.63, 0.84, 1.17	1, 1, 1	0.45, 0.59, 0.77
Hydrogen production	0.63, 0.89, 1.29	1.23, 1.74, 2.39	0.71, 0.99, 1.37	1.3, 1.7, 2.21	1, 1, 1
			CR = 0.0097		

Table 5. Results of possibility degree $S_j = V(S_j \geq S_i)$.

$V(S_1 \geq S_i)$	Value	$V(S_2 \geq S_i)$	Values	$V(S_3 \geq S_i)$	Value	$V(S_4 \geq S_i)$	Value	$V(S_5 \geq S_i)$	Value
$S_1 \geq S_2$	1.00	$S_2 \geq S_1$	0.44	$S_3 \geq S_1$	0.82	$S_4 \geq S_1$	0.55	$S_5 \geq S_1$	0.94
$S_1 \geq S_3$	1.00	$S_2 \geq S_3$	0.63	$S_3 \geq S_2$	1.00	$S_4 \geq S_2$	1.00	$S_5 \geq S_2$	1.00
$S_1 \geq S_4$	1.00	$S_2 \geq S_4$	0.88	$S_3 \geq S_4$	1.00	$S_4 \geq S_3$	0.74	$S_5 \geq S_3$	1.00
$S_1 \geq S_5$	1.00	$S_2 \geq S_5$	0.49	$S_3 \geq S_5$	0.88	$S_4 \geq S_5$	0.60	$S_5 \geq S_4$	1.00

Once values of $V(S_j \geq S_i)$ were compared; we used Equation (10) to find minimum degree possibility $d(i)$ of each variable as below:

$$d'(\text{Capital cost}) = \min(1, 1, 1, 1) = 1.00$$

$$d'(\text{CO}_2 \text{ emission}) = \min(0.442, 0.626, 0.879, 0.489) = 0.44$$

$$d'(\text{Feedstock cost}) = \min(0.817, 1, 1, 0.875) = 0.82$$

$$d'(\text{O\&M}) = \min(0.551, 1, 0.741, 0.603) = 0.55$$

$$d'(\text{Hydrogen production}) = \min(0.94, 1, 1, 1) = 0.94$$

Subsequently, we can define weight vector W' as follows:

$$W' = (1, 0.442, 0.817, 0.551, 0.939)^T$$

Finally, the weight vector W' was normalized using Equation (13) to obtain the initial weights of each criterion. Figure 3 shows the initial weights of criteria.

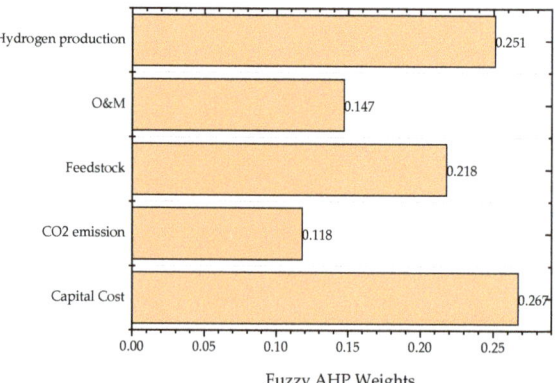

Figure 3. Initial weights of criteria.

4.3. Determine Ultimate Weights Using FTOPSIS

After computing the initial weights of variables, the FTOPSIS was applied to determine the ultimate weights of each variable against each criterion. Firstly, the fuzzy decision matrix was constructed to transform the linguistic variables using Equation (14). Later, Equation (15) was applied to convert fuzzy decision matrix into normalized decision matrix. The normalized decision matrix was then multiplied with FAHP weights to form a weighted decision matrix using Equation (17). Tables 6–8 present fuzzy decision, normalized decision, and weighted decision matrices, respectively.

Table 6. Fuzzy decision matrix of criteria.

	Capital Cost	CO_2 Emission	Feedstock Cost	O&M Cost	Hydrogen Production
Coal gasification	3.3, 4.3, 5.3	6.5, 7.5, 8.5	3, 4, 5	3.5, 4.5, 5.5	5.3, 6.3, 7.3
Natural gas reforming	3.5, 4.5, 5.5	5.5, 6.5, 7.5	3.5, 4.5, 5.5	4.7, 5.7, 6.7	5.7, 6.7, 7.7
Biomass gasification	3.1, 4.1, 5.1	5.6, 6.6, 7.6	4.9, 5.9, 6.9	4.6, 5.6, 6.6	6, 7, 8
Solar thermochemical	4.9, 5.9, 6.9	2, 3, 4	2, 3, 4	3, 4, 5	4.4, 5.4, 6.4
Biomass-derived liquid reforming	4.3, 5.3, 6.3	4.3, 5.3, 6.3	5.2, 6.2, 7.2	4.3, 5.3, 6.3	4.1, 5.1, 6.1
Grid electrolysis	2.9, 3.9, 4.9	5.8, 6.8, 7.8	4.7, 5.7, 6.7	1.5, 2.5, 3.5	6.4, 7.4, 8.4
Wind power electrolysis	3.6, 4.6, 5.6	1.6, 2.6, 3.6	1.7, 2.7, 3.7	2.4, 3.4, 4.4	6.2, 7.2, 8.2
PV electrolysis	3.2, 4.2, 5.2	1.7, 2.7, 3.7	1.5, 2.5, 3.5	2, 3, 4	6, 7, 8
Photoelectrochemical	5.3, 6.3, 7.3	2, 3, 4	4.2, 5.2, 6.2	4.1, 5.1, 6.1	3.8, 4.8, 5.8
Microbial biomass conversion	5.4, 6.4, 7.4	3, 4, 5	4.9, 5.9, 6.9	4.9, 5.9, 6.9	3.9, 4.9, 5.9
Photobiological	4.9, 5.9, 6.9	2.2, 3.2, 4.2	5.7, 6.7, 7.7	4.8, 5.8, 6.8	4.2, 5.2, 6.2

Table 7. Normalized fuzzy decision matrix.

	Capital Cost	CO_2 Emission	Feedstock Cost	O&M Cost	Hydrogen Production
Coal gasification	0.45, 0.58, 0.72	0.76, 0.88, 1	0.39, 0.52, 0.65	0.51, 0.65, 0.8	0.63, 0.75, 0.87
Natural gas reforming	0.47, 0.61, 0.74	0.65, 0.76, 0.88	0.45, 0.58, 0.71	0.68, 0.83, 0.97	0.68, 0.8, 0.92
Biomass gasification	0.42, 0.55, 0.69	0.66, 0.78, 0.89	0.64, 0.77, 0.9	0.67, 0.81, 0.96	0.71, 0.83, 0.95
Solar thermochemical	0.66, 0.8, 0.93	0.24, 0.35, 0.47	0.26, 0.39, 0.52	0.43, 0.58, 0.72	0.52, 0.64, 0.76
Biomass-derived liquid reforming	0.58, 0.72, 0.85	0.51, 0.62, 0.74	0.68, 0.81, 0.94	0.62, 0.77, 0.91	0.49, 0.61, 0.73
Grid electrolysis	0.39, 0.53, 0.66	0.68, 0.8, 0.92	0.61, 0.74, 0.87	0.22, 0.36, 0.51	0.76, 0.88, 1
Wind power electrolysis	0.49, 0.62, 0.76	0.19, 0.31, 0.42	0.22, 0.35, 0.48	0.35, 0.49, 0.64	0.74, 0.86, 0.98
PV electrolysis	0.43, 0.57, 0.7	0.2, 0.32, 0.44	0.19, 0.32, 0.45	0.29, 0.43, 0.58	0.71, 0.83, 0.95
Photoelectrochemical	0.72, 0.85, 0.99	0.24, 0.35, 0.47	0.55, 0.68, 0.81	0.59, 0.74, 0.88	0.45, 0.57, 0.69
Microbial biomass conversion	0.73, 0.86, 1	0.35, 0.47, 0.59	0.64, 0.77, 0.9	0.71, 0.86, 1	0.46, 0.58, 0.7
Photobiological	0.66, 0.8, 0.93	0.26, 0.38, 0.49	0.74, 0.87, 1	0.7, 0.84, 0.99	0.5, 0.62, 0.74

Table 8. Weighted fuzzy decision matrix.

	Capital Cost	CO_2 Emission	Feedstock Cost	O&M Cost	Hydrogen Production
Coal gasification	0.12, 0.16, 0.19	0.09, 0.1, 0.12	0.08, 0.11, 0.14	0.07, 0.1, 0.12	0.16, 0.19, 0.22
Natural gas reforming	0.13, 0.16, 0.2	0.08, 0.09, 0.1	0.1, 0.13, 0.16	0.1, 0.12, 0.14	0.17, 0.2, 0.23
Biomass gasification	0.11, 0.15, 0.18	0.08, 0.09, 0.11	0.14, 0.17, 0.2	0.1, 0.12, 0.14	0.18, 0.21, 0.24
Solar thermochemical	0.18, 0.21, 0.25	0.03, 0.04, 0.06	0.06, 0.08, 0.11	0.06, 0.09, 0.11	0.13, 0.16, 0.19
Biomass-derived liquid reforming	0.16, 0.19, 0.23	0.06, 0.07, 0.09	0.15, 0.18, 0.2	0.09, 0.11, 0.13	0.12, 0.15, 0.18
Grid electrolysis	0.1, 0.14, 0.18	0.08, 0.09, 0.11	0.13, 0.16, 0.19	0.03, 0.05, 0.07	0.19, 0.22, 0.25
Wind power electrolysis	0.13, 0.17, 0.2	0.02, 0.04, 0.05	0.05, 0.08, 0.1	0.05, 0.07, 0.09	0.19, 0.22, 0.25
PV electrolysis	0.12, 0.15, 0.19	0.02, 0.04, 0.05	0.04, 0.07, 0.1	0.04, 0.06, 0.09	0.18, 0.21, 0.24
Photoelectrochemical	0.19, 0.23, 0.26	0.03, 0.04, 0.06	0.12, 0.15, 0.18	0.09, 0.11, 0.13	0.11, 0.14, 0.17
Microbial biomass conversion	0.19, 0.23, 0.27	0.04, 0.06, 0.07	0.14, 0.17, 0.2	0.1, 0.13, 0.15	0.12, 0.15, 0.18
Photobiological	0.18, 0.21, 0.25	0.03, 0.04, 0.06	0.16, 0.19, 0.22	0.1, 0.12, 0.14	0.13, 0.16, 0.19

After constructing the weighted fuzzy decision matrix, FPIS and FNIS were calculated using Equations (18) and (19), respectively. Equation (20) was used to determine the distance of alternatives for each variable from FPIS and FNIS. Equation (21) was applied to obtain the ultimate weights of variables for each alternative. Table A1, Table A2, Table A3, Table A4, Table A5 in Appendix A show values of distance from FPIS and FNIS, and ultimate weights of capital cost, CO_2 emission, feedstock cost, O&M cost, and hydrogen production, respectively. Finally, the ultimate weights were normalized, which are given in Table 9.

Table 9. Normalized ultimate weights.

Process	Technologies	Capital Cost	CO_2 Emission	Feedstock Cost	O&M Cost	Hydrogen Production
Thermochemical process	Coal gasification	0.032	0.217	0.061	0.086	0.106
	Natural gas reforming	0.048	0.173	0.081	0.137	0.134
	Biomass gasification	0.016	0.177	0.137	0.133	0.155
	Solar thermochemical	0.16	0.018	0.02	0.064	0.042
	Biomass-derived liquid reforming	0.112	0.12	0.149	0.12	0.021
Electrolysis	Grid electrolysis	0.16	0.186	0.129	0.039	0.183
	Wind power electrolysis	0.056	0.005	0.008	0.039	0.169
	PV electrolysis	0.024	0.005	0.008	0.021	0.155
Direct solar water splitting process	Photoelectrochemical	0.192	0.018	0.109	0.112	0.007
Biological process	Microbial biomass conversion	0.2	0.062	0.137	0.146	0.007
	Photobiological	0.16	0.027	0.169	0.142	0.028

4.4. Slack-Based Environmental DEA

Slack-based environmental DEA was applied to obtain the directional efficiency of the alternatives. Capital, feedstock, and O&M costs were used as input variables, hydrogen production as desirable output, and CO_2 emission as undesirable output. The ultimate weights of these variables were used in DEA to compute the efficiency scores (given in Table 10).

Table 10. Efficiency scores of alternatives.

Technologies	β	Ranking
Coal gasification	0.9571	8
Natural gas reforming	0.9519	7
Biomass gasification	0	1
Solar thermochemical	0.8708	5
Biomass-derived liquid reforming	0.9897	10
Grid electrolysis	0.573	4
Wind power electrolysis	0	1
PV electrolysis	0	1
Photoelectrochemical	0.9773	9
Microbial biomass conversion	0.9933	11
Photobiological	0.9405	6

The 0 efficiency score implies that the hydrogen production process is fully efficient. Any score above 0 shows inefficiencies in the process. Accordingly, the ranking of hydrogen production processes was undertaken, as shown in Table 9.

The slacks show any shortfall of desirable output, and any excess of inputs and undesirable output, as given in Table 11.

Table 11. Slacks of inputs, desirable outputs, and undesirable outputs.

Alternatives	Inputs (Excess)						Desirable Output (Shortfall)		Undesirable Output (Excess)	
	Capital Cost	Slack (%)	Feedstock Cost	Slack (%)	O&M Cost	Slack (%)	Hydrogen Production	Slack (%)	CO_2 Emission	Slack (%)
Coal gasification	0	0	0.0483	−79.18	0.0562	−65.35	0	0	0.0813	−37.47
Natural gas reforming	0	0	0.0677	−83.58	0.0975	−71.17	0	0	0.0507	−29.31
Biomass gasification	0	0	0	0	0	0	0	0	0	0
Solar thermochemical	0.134	−83.75	0.0163	−81.5	0.0459	−71.72	0	0	0	0
Biomass-derived liquid reforming	0.0982	−87.68	0.147	−98.66	0.1104	−92	0	0	0	0
Grid electrolysis	0.1154	−72.13	0.1141	−88.45	0	0	0	0	0.0701	−37.69
Wind power electrolysis	0	0	0	0	0	0	0	0	0	0
PV electrolysis	0	0	0	0	0	0	0	0	0	0
Photoelectrochemical	0.1874	−97.6	0.1083	−99.36	0.1088	−97.14	0	0	0	0
Microbial biomass conversion	0.1954	−97.7	0.1363	−99.49	0.1428	−97.81	0	0	0	0
Photobiological	0.142	−88.75	0.1664	−98.46	0.1295	−91.2	0	0	0	0

5. Results and Findings

The results of the study are divided into three parts. The first part of the results comprises initial weights, calculated using FAHP, the second presents ultimate weights, computed using FTOPSIS, while the final part presents the ranking of alternatives and analyses of inputs, desirable output, and undesirable output slacks, measured using slack-based environmental DEA.

The result of FAHP shows that the capital cost, which is an input variable, received the highest initial weight of 0.267. Hydrogen production, which is a desirable output variable, achieved the second highest weight of 0.251. Feedstock, an input variable, got the third highest weight of 0.218. CO_2 emission, an undesirable output, received 0.118 while O&M, an input variable, reported achieving the lowest initial weight of 0.147.

We used these initial weights in FTOPSIS to compute the ultimate weights of criteria for each alternative. By doing so, we could also minimize the vagueness involved in the process of obtaining criteria weights. Table 1 presents the results of the ultimate weights calculated for each alternative.

Finally, the slack-based environmental DEA ranks the alternatives, i.e., hydrogen production technologies, according to their feasibility in the context of Pakistan. Table 1 shows the final ranking of alternatives. It can be seen that three technologies, i.e., PV electrolysis, wind power electrolysis, and biomass gasification, received an efficiency score of 0, which shows the level of full efficiency. Subsequently, these three technologies were ranked first. The interesting point here is that all of these three technologies produce hydrogen from renewable energy sources, which are abundant in Pakistan. Additionally, these technologies are mature and already being used to generate electricity in the country.

Grid electrolysis achieved the fourth position. However, it is still not suggested due to being an inefficient source as we can see its efficiency score is 0.573, which is quite larger than an efficient score of '0'. The rest of the ranking is as follows: Solar thermochemical > Photobiological > Natural gas reforming > Coal gasification > Photoelectrochemical > Biomass-derived liquid reforming > Microbial biomass conversion.

Currently, except first-ranked technologies, all the rest are not recommended in Pakistan. To meet the environmental efficiency level, set by the top-three technologies, the rest of the technologies need a massive reduction, mainly in their input variables. The slack analysis enables to find out necessary reductions in inputs, and undesirable outputs. The results of the slack analysis, given in Table 2, show that in order to achieve a fully efficient level, the grid electrolysis must reduce capital cost by 72.13%, feedstock cost by 88.45%, and CO_2 emission by 37.69%.

Similarly, the photobiological technology can be efficient if capital cost is reduced by 88.75%, feedstock cost by 98.46%, and O&M cost by 91.20%. Natural gas reforming must decrease feedstock cost by 83.58%, and O&M cost by 71.17%. Coal gasification needs to reduce feedstock, O&M, and CO_2 emission by 79.18%, 65.35%, and 37.47%, respectively. For photoelectrochemical to achieve an efficient level, there is a need to decrease capital, feedstock, and O&M costs by 97.60%, 99.36%, and 97.14%, respectively. Biomass-derived liquid reforming can be an efficient hydrogen production option in Pakistan if the technology reduces capital cost by 87.68%, feedstock cost by 98.66%, and O&M cost by 92%. The microbial biomass conversion technology needs 97.70% reduction in capital cost, 99.49% reduction in feedstock cost, and 97.81% reduction in O&M cost.

6. Conclusions

The paper presented a framework to evaluate the economic and environmental efficiency of hydrogen production processes for decarbonization of energy systems. Since the production processes produce undesirable outputs as well, therefore, environmental DEA was applied for assessing the sustainability of these processes. A common problem that arises from applying environmental DEA is that the DEA ignores the relative importance of variables while assigning weights to each variable. Tackling this issue, the proposed framework firstly used two widely-applied MCDA techniques, i.e., FAHP and FTOPSIS, before employing the environmental DEA to assess the efficiency of hydrogen production processes.

The proposed framework was applied to prioritize the most sustainable hydrogen production process in Pakistan. Eleven hydrogen production alternatives under four main processes, i.e., thermochemical, electrolysis, direct solar water splitting, and biological processes, were analyzed. Five alternatives under the thermochemical process include coal gasification, biomass gasification, solar thermochemical, natural gas reforming, and biomass-derived liquid reforming. The electrolysis processes included three alternatives, i.e., grid electrolysis, wind electrolysis, and PV electrolysis. The photoelectrochemical alternative was selected under the direct solar water splitting process, whereas microbial biomass conversion and photobiological alternatives were shortlisted for analysis under the biological process.

Shortlisted alternatives were evaluated based on five criteria. These criteria included three inputs (capital cost, O&M cost, and feedstock cost), one desirable output (hydrogen production), and one undesirable output (CO_2 emission). The initial weights of criteria were obtained using FAHP, and then FTOPSIS was applied to compute the ultimate weights of each criterion for each alternative. Finally, the slack-based environmental DEA was employed to assess the most sustainable hydrogen production process in the Pakistan context. The results of the study showed that the wind electrolysis, PV electrolysis, and biomass gasification are the most sustainable hydrogen production processes in Pakistan. The rest of the eight alternatives were not recommended due to their poor efficiency scores. However, these may become sustainable choices in the future if deficiencies pointed out in slack analysis are appropriately improved.

Author Contributions: All the authors contributed to this work. L.X. and Y.W. supervised the work and developed the concept. S.A.A.S. prepared the original draft and structured the study. S.A.A.S., H.Z. and Y.A.S. undertook the survey and developed the model and preliminary manuscript. G.D.W. and Z.A.S. analyzed the results and finalized the manuscript.

Funding: 1. National Natural Science Foundation of China (Grant No. 71873064), Research on OFDI driving low-carbon upgrading of China's equipment manufacturing global value chain: theoretical mechanism, implementation path and performance evaluation. 2. General Projects of Humanities and Social Sciences of the Ministry of Education (Planning Projects) (Grant No. 18YJA790085), Performance evaluation of OFDI driving low-carbon upgrading of China's equipment.

Conflicts of Interest: The authors declare no conflict of interest.

Appendix A

Table A1. Distance from fuzzy positive ideal solution (FPIS) and fuzzy negative ideal solution (FNIS), and ultimate weights of capital cost.

	Capital Cost		
	d^+	d^-	Ultimate Weight
Coal gasification	1.0499	0.2008	0.1606
Natural gas reforming	0.9501	0.3006	0.2403
Biomass gasification	1.1496	0.0997	0.0798
Solar thermochemical	0.2493	1	0.8004
Biomass-derived liquid reforming	0.5499	0.7004	0.5602
Grid electrolysis	1.1496	0.0997	0.0798
Wind power electrolysis	0.8989	0.3504	0.2805
PV electrolysis	1.0997	0.1498	0.1199
Photoelectrochemical	0.0499	1.1994	0.9601
Microbial biomass conversion	0	1.2493	1
Photobiological	0.2493	1	0.8004

Table A2. Distance from FPIS and FNIS, and ultimate weights of CO_2 emission.

	CO_2 Emission		
	d+	d−	Ultimate Weight
Coal gasification	0	2.446	1
Natural gas reforming	0.5	1.9531	0.7962
Biomass gasification	0.4496	1.9964	0.8162
Solar thermochemical	2.2482	0.1986	0.0812
Biomass-derived liquid reforming	1.0971	1.3489	0.5515
Grid electrolysis	0.3489	2.0971	0.8574
Wind power electrolysis	2.3957	0.0504	0.0206
PV electrolysis	2.3957	0.0504	0.0206
Photoelectrochemical	2.2482	0.1986	0.0812
Microbial biomass conversion	1.7482	0.6978	0.2853
Photobiological	2.1475	0.2986	0.1221

Table A3. Distance from FPIS and FNIS, and ultimate weights of feedstock cost.

	Feedstock Cost		
	d+	d−	Ultimate Weight
Coal gasification	1.3498	0.7496	0.3571
Natural gas reforming	1.1007	1	0.476
Biomass gasification	0.4011	1.6996	0.8091
Solar thermochemical	1.8483	0.2491	0.1188
Biomass-derived liquid reforming	0.2504	1.8498	0.8808
Grid electrolysis	0.5	1.6007	0.762
Wind power electrolysis	2.0018	0.0989	0.0471
PV electrolysis	2.0018	0.0989	0.0471
Photoelectrochemical	0.7509	1.3498	0.6425
Microbial biomass conversion	0.4011	1.6996	0.8091
Photobiological	0	2.1007	1

Table A4. Distance from FPIS and FNIS, and ultimate weights of O&M Cost.

	O&M Cost		
	d+	d−	Ultimate Weight
Coal gasification	0.6995	1	0.5884
Natural gas reforming	0.1009	1.5986	0.9406
Biomass gasification	0.1502	1.5493	0.9116
Solar thermochemical	0.9507	0.7488	0.4406
Biomass-derived liquid reforming	0.3005	1.3991	0.8232
Grid electrolysis	1.2512	0.4484	0.2638
Wind power electrolysis	1.2512	0.4484	0.2638
PV electrolysis	1.4507	0.2488	0.1464
Photoelectrochemical	0.4005	1.2981	0.7642
Microbial biomass conversion	0	1.6995	1
Photobiological	0.0493	1.6502	0.971

Table A5. Distance from FPIS and FNIS, and ultimate weights of hydrogen production.

	Hydrogen Production		
	d^+	d^-	Ultimate Weight
Coal gasification	0.5494	0.7508	0.5774
Natural gas reforming	0.3495	0.9498	0.731
Biomass gasification	0.199	1.1003	0.8468
Solar thermochemical	1	0.301	0.2314
Biomass-derived liquid reforming	1.1488	0.1505	0.1158
Grid electrolysis	0	1.2993	1
Wind power electrolysis	0.1003	1.201	0.9229
PV electrolysis	0.199	1.1003	0.8468
Photoelectrochemical	1.2492	0.0502	0.0386
Microbial biomass conversion	1.2492	0.0502	0.0386
Photobiological	1.0987	0.2007	0.1545

References

1. Parra, D.; Valverde, L.; Pino, F.J.; Patel, M.K. A Review on the Role, Cost and Value of Hydrogen Energy Systems for Deep Decarbonisation. *Renew. Sustain. Energy Rev.* **2019**, *101*, 279–294. [CrossRef]
2. Staffell, I.; Scamman, D.; Abad, A.V.; Balcombe, P.; Dodds, P.E.; Ekins, P.; Shah, N.; Ward, K.R. The Role of Hydrogen and Fuel Cells in the Global Energy System. *Energy Environ. Sci.* **2019**, *12*, 463–491. [CrossRef]
3. Hanley, E.S.; Deane, J.P.; Gallachóir, B.P.Ó. The Role of Hydrogen in Low Carbon Energy Futures—A Review of Existing Perspectives. *Renew. Sustain. Energy Rev.* **2018**, *82*, 3027–3045. [CrossRef]
4. Grochala, W. First There Was Hydrogen. *Nat. Chem.* **2015**, *7*, 264. [CrossRef] [PubMed]
5. Dincer, I.; Acar, C. A Review on Clean Energy Solutions for Better Sustainability. *Int. J. Energy Res.* **2015**, *39*, 585–606. [CrossRef]
6. Cooperberg, D. Industrial Applications of Hydrogen. *Hydrog. Technol. Implic. Util. Hydrog.* **2018**, *4*.
7. Melton, N.; Axsen, J.; Sperling, D. Moving beyond Alternative Fuel Hype to Decarbonize Transportation. *Nat. Energy* **2016**, *1*, 16013. [CrossRef]
8. Dodds, P.E.; Staffell, I.; Hawkes, A.D.; Li, F.; Grünewald, P.; McDowall, W.; Ekins, P. Hydrogen and Fuel Cell Technologies for Heating: A Review. *Int. J. Hydrog. Energy* **2015**, *40*, 2065–2083. [CrossRef]
9. Eichman, J.; Townsend, A.; Melaina, M. *Economic Assessment of Hydrogen Technologies Participating in California Electricity Markets*; National Renewable Energy Lab.(NREL): Golden, CO, USA, 2016.
10. Pudukudy, M.; Yaakob, Z.; Mohammad, M.; Narayanan, B.; Sopian, K. Renewable Hydrogen Economy in Asia–Opportunities and Challenges: An Overview. *Renew. Sustain. Energy Rev.* **2014**, *30*, 743–757. [CrossRef]
11. Midilli, A.; Dincer, I. Hydrogen as a Renewable and Sustainable Solution in Reducing Global Fossil Fuel Consumption. *Int. J. Hydrog. Energy* **2008**, *33*, 4209–4222. [CrossRef]
12. Sørensen, B.; Spazzafumo, G. *Hydrogen and Fuel Cells: Emerging Technologies and Applications*; Academic Press: Cambridge, MA, USA, 2018.
13. Nikolaidis, P.; Poullikkas, A. A Comparative Overview of Hydrogen Production Processes. *Renew. Sustain. Energy Rev.* **2017**, *67*, 597–611. [CrossRef]
14. Coelli, T.J.; Rao, D.S.P.; O'Donnell, C.J.; Battese, G.E. *An Introduction to Efficiency and Productivity Analysis*; Springer Science & Business Media: Berlin, Germany, 2005.
15. Zimmermann, H.-J. *Fuzzy Set Theory—And Its Applications*; Springer Science & Business Media: Berlin, Germany, 2011.
16. Zhou, P.; Ang, B.W.; Poh, K.L. Slacks-Based Efficiency Measures for Modeling Environmental Performance. *Ecol. Econ.* **2006**, *60*, 111–118. [CrossRef]
17. Acar, C.; Beskese, A.; Temur, G.T. Sustainability Analysis of Different Hydrogen Production Options Using Hesitant Fuzzy AHP. *Int. J. Hydrog. Energy* **2018**, *43*, 18059–18076. [CrossRef]
18. Ren, J.; Toniolo, S. Life Cycle Sustainability Decision-Support Framework for Ranking of Hydrogen Production Pathways under Uncertainties: An Interval Multi-Criteria Decision Making Approach. *J. Clean. Prod.* **2018**, *175*, 222–236. [CrossRef]

19. Ren, J.; Gao, S.; Tan, S.; Dong, L.; Scipioni, A.; Mazzi, A. Role Prioritization of Hydrogen Production Technologies for Promoting Hydrogen Economy in the Current State of China. *Renew. Sustain. Energy Rev.* **2015**, *41*, 1217–1229. [CrossRef]
20. Yu, D. Hydrogen Production Technologies Evaluation Based on Interval-Valued Intuitionistic Fuzzy Multiattribute Decision Making Method. *J. Appl. Math.* **2014**, *2014*. [CrossRef]
21. Ren, J.; Fedele, A.; Mason, M.; Manzardo, A.; Scipioni, A. Fuzzy Multi-Actor Multi-Criteria Decision Making for Sustainability Assessment of Biomass-Based Technologies for Hydrogen Production. *Int. J. Hydrog. Energy* **2013**, *38*, 9111–9120. [CrossRef]
22. Pilavachi, P.A.; Chatzipanagi, A.I.; Spyropoulou, A.I. Evaluation of Hydrogen Production Methods Using the Analytic Hierarchy Process. *Int. J. Hydrog. Energy* **2009**, *34*, 5294–5303. [CrossRef]
23. Saaty, T.L. Analytic Heirarchy Process. *Wiley Statsref Stat. Ref. Online* **2014**. [CrossRef]
24. Shah, S.A.A.; Solangi, Y.A.; Ikram, M. Analysis of Barriers to the Adoption of Cleaner Energy Technologies in Pakistan Using Modified Delphi and Fuzzy Analytical Hierarchy Process. *J. Clean. Prod.* **2019**, *235*, 1037–1050. [CrossRef]
25. Patil, S.K.; Kant, R. A Fuzzy AHP-TOPSIS Framework for Ranking the Solutions of Knowledge Management Adoption in Supply Chain to Overcome Its Barriers. *Expert Syst. Appl.* **2014**, *41*, 679–693. [CrossRef]
26. Chang, D.-Y. Applications of the Extent Analysis Method on Fuzzy AHP. *Eur. J. Oper. Res.* **1996**, *95*, 649–655. [CrossRef]
27. Lai, Y.-J.; Liu, T.-Y.; Hwang, C.-L. Topsis for MODM. *Eur. J. Oper. Res.* **1994**, *76*, 486–500. [CrossRef]
28. Şengül, Ü.; Eren, M.; Shiraz, S.E.; Gezder, V.; Şengül, A.B. Fuzzy TOPSIS Method for Ranking Renewable Energy Supply Systems in Turkey. *Renew. Energy* **2015**, *75*, 617–625. [CrossRef]
29. Solangi, Y.; Tan, Q.; Khan, M.; Mirjat, N.; Ahmed, I. The Selection of Wind Power Project Location in the Southeastern Corridor of Pakistan: A Factor Analysis, AHP, and Fuzzy-TOPSIS Application. *Energies* **2018**, *11*, 1940. [CrossRef]
30. Shah, S.A.A.; Zhou, P.; Walasai, G.D.; Mohsin, M. Energy Security and Environmental Sustainability Index of South Asian Countries: A Composite Index Approach. *Ecol. Indic.* **2019**, *106*, 105507. [CrossRef]
31. Wang, Y.; Shah, S.A.A.; Zhou, P. City-Level Environmental Performance in China. *Energy Ecol. Environ.* **2018**, *3*, 149–161. [CrossRef]
32. Charnes, A.; Cooper, W.W.; Rhodes, E. Measuring the Efficiency of Decision Making Units. *Eur. J. Oper. Res.* **1978**, *2*, 429–444. [CrossRef]
33. Koopmans, T.C. An Analysis of Production as an Efficient Combination of Activities. *Act. Anal. Prod. Alloc.* **1951**.
34. Chung, Y.H.; Färe, R.; Grosskopf, S. Productivity and Undesirable Outputs: A Directional Distance Function Approach. *J. Environ. Manag.* **1997**, *51*, 229–240. [CrossRef]
35. Cooper, W.W.; Seiford, L.M.; Tone, K. *Introduction to Data Envelopment Analysis and Its Uses: With DEA-Solver Software and References*; Springer Science & Business Media: Berlin, Germany, 2006.
36. Shah, S.A.A.; Valasai, G.D.; Memon, A.A.; Laghari, A.N.; Jalbani, N.B.; Strait, J.L. Techno-Economic Analysis of Solar PV Electricity Supply to Rural Areas of Balochistan, Pakistan. *Energies* **2018**, *11*, 1777. [CrossRef]
37. U.S. Department of Energy. Hydrogen Production and Distribution. Available online: https://afdc.energy.gov/fuels/hydrogen_production.html (accessed on 10 May 2019).
38. Baykara, S.Z. Hydrogen: A Brief Overview on Its Sources, Production and Environmental Impact. *Int. J. Hydrog. Energy* **2018**, *43*, 10605–10614. [CrossRef]
39. Gondal, I.A.; Masood, S.A.; Khan, R. Green Hydrogen Production Potential for Developing a Hydrogen Economy in Pakistan. *Int. J. Hydrog. Energy* **2018**, *43*, 6011–6039. [CrossRef]
40. Seyitoglu, S.S.; Dincer, I.; Kilicarslan, A. Energy and Exergy Analyses of Hydrogen Production by Coal Gasification. *Int. J. Hydrog. Energy* **2017**, *42*, 2592–2600. [CrossRef]
41. Shayan, E.; Zare, V.; Mirzaee, I. Hydrogen Production from Biomass Gasification; a Theoretical Comparison of Using Different Gasification Agents. *Energy Convers. Manag.* **2018**, *159*, 30–41. [CrossRef]
42. Rao, C.N.R.; Dey, S. Solar Thermochemical Splitting of Water to Generate Hydrogen. *Proc. Natl. Acad. Sci. USA* **2017**, *114*, 13385–13393. [CrossRef]
43. Villafán-Vidales, H.I.; Arancibia-Bulnes, C.A.; Riveros-Rosas, D.; Romero-Paredes, H.; Estrada, C.A. An Overview of the Solar Thermochemical Processes for Hydrogen and Syngas Production: Reactors, and Facilities. *Renew. Sustain. Energy Rev.* **2017**, *75*, 894–908. [CrossRef]

44. Garcia, L.; French, R.; Czernik, S.; Chornet, E. Catalytic Steam Reforming of Bio-Oils for the Production of Hydrogen: Effects of Catalyst Composition. *Appl. Catal. A Gen.* **2000**, *201*, 225–239. [CrossRef]
45. Schmitt, N.; Apfelbacher, A.; Jäger, N.; Daschner, R.; Stenzel, F.; Hornung, A. Thermo-chemical Conversion of Biomass and Upgrading to Biofuel: The Thermo-Catalytic Reforming Process—A Review. *Biofuels Bioprod. Biorefin.* **2019**, *13*, 822–837. [CrossRef]
46. da Silva Veras, T.; Mozer, T.S.; da Silva César, A. Hydrogen: Trends, Production and Characterization of the Main Process Worldwide. *Int. J. Hydrog. Energy* **2017**, *42*, 2018–2033. [CrossRef]
47. Mallouk, T.E. Water Electrolysis: Divide and Conquer. *Nat. Chem.* **2013**, *5*, 362. [CrossRef]
48. Xu, L.; Wang, Y.; Solangi, Y.A.; Zameer, H.; Shah, S.A.A. Off-Grid Solar PV Power Generation System in Sindh, Pakistan: A Techno-Economic Feasibility Analysis. *Processes* **2019**, *7*, 308. [CrossRef]
49. Saba, S.M.; Müller, M.; Robinius, M.; Stolten, D. The Investment Costs of Electrolysis—A Comparison of Cost Studies from the Past 30 Years. *Int. J. Hydrog. Energy* **2018**, *43*, 1209–1223. [CrossRef]
50. Li, Z.; Guo, P.; Han, R.; Sun, H. Current Status and Development Trend of Wind Power Generation-Based Hydrogen Production Technology. *Energy Explor. Exploit.* **2019**, *37*, 5–25. [CrossRef]
51. Chi, J.; Yu, H. Water Electrolysis Based on Renewable Energy for Hydrogen Production. *Chin. J. Catal.* **2018**, *39*, 390–394. [CrossRef]
52. Badwal, S.P.S.; Giddey, S.; Munnings, C. Emerging Technologies, Markets and Commercialization of Solid-electrolytic Hydrogen Production. *Wiley Interdiscip. Rev. Energy Environ.* **2018**, *7*, e286. [CrossRef]
53. Van de Krol, R.; Grätzel, M. *Photoelectrochemical Hydrogen Production*; Springer: New York, NY, USA, 2012; Volume 90.
54. Singh, L.; Wahid, Z.A. Methods for Enhancing Bio-Hydrogen Production from Biological Process: A Review. *J. Ind. Eng. Chem.* **2015**, *21*, 70–80. [CrossRef]
55. Das, D.; Veziroğlu, T.N. Hydrogen Production by Biological Processes: A Survey of Literature. *Int. J. Hydrog. Energy* **2001**, *26*, 13–28. [CrossRef]
56. Kumar, G.; Bakonyi, P.; Kobayashi, T.; Xu, K.-Q.; Sivagurunathan, P.; Kim, S.-H.; Buitrón, G.; Nemestóthy, N.; Bélafi-Bakó, K. Enhancement of Biofuel Production via Microbial Augmentation: The Case of Dark Fermentative Hydrogen. *Renew. Sustain. Energy Rev.* **2016**, *57*, 879–891. [CrossRef]
57. Kumar, G.; Shobana, S.; Nagarajan, D.; Lee, D.-J.; Lee, K.-S.; Lin, C.-Y.; Chen, C.-Y.; Chang, J.-S. Biomass Based Hydrogen Production by Dark Fermentation—Recent Trends and Opportunities for Greener Processes. *Curr. Opin. Biotechnol.* **2018**, *50*, 136–145. [CrossRef]
58. Cabrol, L.; Marone, A.; Tapia-Venegas, E.; Steyer, J.-P.; Ruiz-Filippi, G.; Trably, E. Microbial Ecology of Fermentative Hydrogen Producing Bioprocesses: Useful Insights for Driving the Ecosystem Function. *Fems Microbiol. Rev.* **2017**, *41*, 158–181. [CrossRef]
59. Singh, R.; White, D.; Demirel, Y.; Kelly, R.; Noll, K.; Blum, P. Uncoupling Fermentative Synthesis of Molecular Hydrogen from Biomass Formation in Thermotoga Maritima. *Appl. Environ. Microbiol.* **2018**, *84*, e00998-18. [CrossRef]
60. Khetkorn, W.; Rastogi, R.P.; Incharoensakdi, A.; Lindblad, P.; Madamwar, D.; Pandey, A.; Larroche, C. Microalgal Hydrogen Production—A Review. *Bioresour. Technol.* **2017**, *243*, 1194–1206. [CrossRef]

© 2019 by the authors. Licensee MDPI, Basel, Switzerland. This article is an open access article distributed under the terms and conditions of the Creative Commons Attribution (CC BY) license (http://creativecommons.org/licenses/by/4.0/).

Article

Syngas Production from Combined Steam Gasification of Biochar and a Sorption-Enhanced Water–Gas Shift Reaction with the Utilization of CO_2

Supanida Chimpae [1], Suwimol Wongsakulphasatch [1], Supawat Vivanpatarakij [2,*], Thongchai Glinrun [3], Fasai Wiwatwongwana [4], Weerakanya Maneeprakorn [5] and Suttichai Assabumrungrat [6]

[1] Department of Chemical Engineering, Faculty of Engineering, King Mongkut's University of Technology North Bangkok, Bangkok 10800, Thailand; s5601031620023@email.kmutnb.ac.th (S.C.); suwimol.w@eng.kmutnb.ac.th (S.W.)
[2] Energy Research Institute, Chulalongkorn University, Phayathai Road, Wang Mai, Phatumwan, Bangkok 10330, Thailand
[3] Department of Petrochemical and Environmental Engineering, Faculty of Engineering, Pathumwan Institute of Technology, Rama 1 Road, Wang Mai, Phatumwan, Bangkok 10330, Thailand; thongchai@pit.ac.th
[4] Department of Advanced Manufacturing Technology, Faculty of Engineering, Pathumwan Institute of Technology, 833 Rama 1 Road, Wangmai, Pathumwan, Bangkok 10330, Thailand; fasiaw227@gmail.com
[5] National Nanotechnology Center (NANOTEC), National Science and Technology Development Agency (NSTDA), Pathum Thani 12120, Thailand; weerakanya@nanotec.or.th
[6] Center of Excellence in Catalysis and Catalytic Reaction Engineering, Department of Chemical Engineering, Faculty of Engineering, Chulalongkorn University, Bangkok 10330, Thailand; Suttichai.A@chula.ac.th
* Correspondence: supawat.v@chula.ac.th

Received: 24 April 2019; Accepted: 1 June 2019; Published: 7 June 2019

Abstract: This research aims at evaluating the performance of a combined system of biochar gasification and a sorption-enhanced water–gas shift reaction (SEWGS) for synthesis gas production. The effects of mangrove-derived biochar gasification temperature, pattern of combined gasification and SEWGS, amount of steam and CO_2 added as gasifying agent, and SEWGS temperature were studied in this work. The performances of the combined process were examined in terms of biochar conversion, gaseous product composition, and CO_2 emission. The results revealed that the hybrid SEWGS using one-body multi-functional material offered a greater amount of H_2 with a similar amount of CO_2 emissions when compared with separated sorbent/catalyst material. The gasification temperature of 900 °C provided the highest biochar conversion of ca. 98.7%. Synthesis gas production was found to depend upon the amount of water and CO_2 added and SEWGS temperature. Higher amounts of H_2 were observed when increasing the amount of water and the temperature of the SEWGS system.

Keywords: gasification; sorption-enhanced water–gas shift; multi-functional material

1. Introduction

Synthesis gas or syngas, which is composed mainly of H_2 and CO, can be applied for various downstream processes, e.g., electricity generation or chemical production [1–3]. The conversion of biomass by thermochemical processes such as gasification or pyrolysis has been extensively used to produce syngas and is recognized as an environmental-friendly technique as it is carbon-neutral [4]. The thermochemical process can be performed using different operating conditions, i.e., gasifying agent, temperature, pressure, etc., which could yield different amounts and compositions of syngas [5–7]. In addition, strategic techniques have also been applied for upgrading syngas, i.e., integrated gas–solid simultaneous gasification and catalytic reforming [8], a two-stage pyrolysis-reforming system [9], a two-stage gasification-reforming system [10], catalytic pyrolysis of biomass in a two-stage fixed bed reactor system [11], etc. For example, Chaiwatanodom et al. [6] studied the production of syngas from biomass gasification using recycled CO_2 from the process as a gasifying agent by process modelling using the Aspen Plus program. The authors showed that the ratio of syngas production was varied depending upon amount of CO_2 fed into the system, gasification temperature, and pressure. Waheed et al. [12] studied the production of hydrogen from biochar derived from sugar cane bagasse pyrolysis via steam catalytic gasification. Type of catalyst, gasification temperature, and steam flow rate were found to affect hydrogen yield.

Although biomass gasification has been proven to be one of the most efficient techniques for syngas production, one drawback of this technique is the production of CO_2 in the product stream [13–16]. As is known, the release of CO_2 is a cause of the greenhouse gas effect; storage or utilization of CO_2 has therefore attracted interest worldwide. In our previous work [7], utilization of the released CO_2 as a co-gasifying agent has been investigated for combined gasification with the steam reforming process via thermodynamic analysis using the Aspen Plus program. The results showed that the use of CO_2 recycled from a separation process as a co-gasifying agent could enhance coal gas efficiency and reduce CO_2 emissions. However, syngas composition was obtained differently depending upon combination pattern as well as reforming temperature and feed ratio; separation of CO_2 after gasification process offered a higher H_2/CO ratio when compared with the system that extracted CO_2 after the reforming process. Higher reforming temperature and H_2O feed can lead to higher production of H_2. In this work, the combination of biochar gasification and the reforming process for syngas production is experimentally investigated using a packed-bed reactor system. Effects of combination pattern, operating temperature, feed ratio of gasifying agent, and amount of catalyst on syngas production and CO_2 emission are examined. In addition, we have applied the concept of sorption-enhanced steam reforming by using a one-body multi-functional material, which contains CO_2 sorbent and catalyst, to the reforming system with the purpose on improving process efficiency.

2. Materials and Methods

2.1. Material Synthesis

In this work, 12.5 wt.% of Ni on a γ-Al_2O_3 support was used as reforming catalyst, as it has been proven that it is suitable for steam reforming [17]. The material was prepared by the wet impregnation method using $Ni(NO_3)_2 \cdot 6H_2O$ as precursor. Firstly, 6.66 g of $Ni(NO_3)_2 \cdot 6H_2O$ was dissolved in 80 mL of deionized water, then 17.87 g of γ-Al_2O_3 was added into aqueous nickel nitrate solution and stirred at 80 °C until the water was almost completely evaporated. The solid was dried at 120 °C overnight and calcined at 600 °C for 3 h in air.

CaO on Al_2O_3 support, named $CaO/Ca_{12}Al_{14}O_{33}$, was used as CO_2 adsorbent as it offers high CO_2 sorption capacity in the temperature range of steam reforming [18]. In this work, $CaO/Ca_{12}Al_{14}O_{33}$ was synthesized by the sol-gel method using $Al(NO_3)_3 \cdot 9H_2O$ and $Ca(NO_3)_2 \cdot 4H_2O$ as precursors. To prepare this sorbent, 4.22 g of $Ca(NO_3)_2 \cdot 4H_2O$ was mixed with 2.31 g of $Al(NO_3)_3 \cdot 9H_2O$ in DI (deionized) water. Then, 5.02 g of citric acid were added into the solution, which was stirred at 80 °C for 7 h. After that, the mixture was placed at ambient temperature for 18 h to form wet gel. Later, the wet gel was dried at 80 °C for 5 h and at 110 °C for 12 h, respectively, followed by calcination at 850 °C for 2 h under dried air. At this stage, $CaO/Ca_{12}Al_{14}O_{33}$ containing $CaO:Ca_{12}Al_{14}O_{33}$ = 70:30 wt.% was obtained.

One-body multi-functional sorbent/catalyst material, designated as xwt.% NiO/CaO-$Ca_{12}Al_{14}O_{33}$, was prepared by sol-gel method following Changjun et al. [19]. In brief, 3.11 g of $Ni(NO_3)_2 \cdot 6H_2O$, 5.69 g of $Al(NO_3)_3 \cdot 9H_2O$, and 17.81 g of $Ca(NO_3)_2 \cdot 4H_2O$, were dissolved in 109 mL of DI water with the addition of citric acid using a molar ratio of citric acid to Al^{3+}, Ni^{2+}, Ca^{2+} equal to 1.2:1:1:1. The solution was adjusted to pH 1–2 by nitric acid. Then, the solution was heated up and stirred at 80 °C under reflux for 2 h. After that, ethylene glycol (mass ratio polyethylene glycol to citric acid of 0.5) was added into the solution, and stirred under reflux at 105 °C for 5 h. The solution was thereafter dried in an oven at 110 °C for 12 h and calcined at 850 °C for 2 h under dried air.

2.2. Material Characterization

Synthetic materials were characterized their compositions and crystallinity by the X-ray diffraction (XRD) technique; Bruker model D8 Advance (Bruker Crop., Billerica, MA, USA). Surface area, pore size, and pore volume were investigated by N_2 adsorption/desorption isotherm by Brunauer–Emmett–Teller (BET) technique; Micromeritics model 3Flex (Micrometrics Instrument Corp., Norcross, GA, USA). Morphologies of the samples were determined by a scanning electron microscope (SEM); Hitachi model S-3400N (Hitachi High-Technologies Corp., Tokyo, Japan).

2.3. Syngas Production Test

Syngas production experiments were carried out by using two-connected fixed-bed reactors, one for biomass gasification and the other for reforming reaction (see Figure 1). Prior to running experiment, biochar was pretreated by Ar with a flow rate of 50 mL/min at 600 °C for 60 min. Sorbent and catalyst materials were pretreated by Ar with a flow rate of 50 mL/min at 850 °C for 30 min followed by the same flow rate of H_2 at 850 °C for 30 min, respectively. In this work, gasification temperature was varied between 850 °C and 950 °C and that of reforming was varied between 500 °C and 650 °C under atmospheric pressure. The gasifying agent was fed at a fixed ratio of O_2 and C, whereas CO_2 and H_2O were varied. The $CO_2/O_2/H_2O/C$ feed ratios were varied in the range of 0–0.5:0.125:0–1.5:1. All experiments were carried out by fixing total feed flow rate to yield gas hourly space velocity (GHSV) ca. 700 h^{-1}.

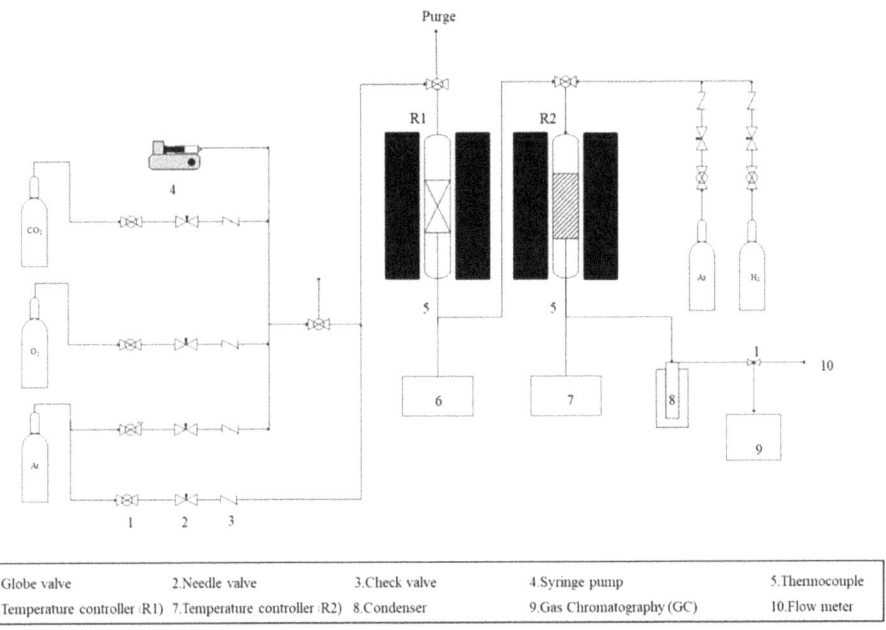

Figure 1. Schematic diagram of experimental setup for syngas production.

Performances of the combined process were determined in terms of biochar conversion (%Biochar conversion), ratio of H_2/CO in the produced syngas (H_2/CO ratio), and CO_2 emission ratio (CO_2 EMR) as defined as follows:

%Biochar conversion:

$$\%\text{Biochar conversion} = \frac{\text{mole of biochar}_{in} - \text{mole of biochar}_{out}}{\text{mole of biochar}_{in}} \times 100 \qquad (1)$$

H_2/CO ratio:

$$H_2/CO \text{ ratio} = \frac{\text{mole of } H_2 \text{ produced}}{\text{mole of CO produced}} \qquad (2)$$

CO_2 emission ratio, CO_2 EMR:

$$CO_2 \text{ emission ratio}(CO_2 \text{ EMR}) = \frac{\text{mole of } CO_2 \text{ emission from } CO_2 \text{outlets}}{\text{mole of } CO_2 \text{ total}} \qquad (3)$$

where CO_2 total is the amount of CO_2 produced from the gasifier.

3. Results and Discussions

3.1. Effect of Gasification Temperature

Conversion of biochar was firstly investigated by studying the effect of gasification temperature using H_2O and O_2 as gasifying agents with a $H_2O:O_2:C$ feed molar ratio of 0.25:0.25:1. As shown in Figure 2, high biochar conversions of 97.5%, 98.7%, and 98.1% could be obtained by gasification at temperatures of 850 °C, 900 °C, and 950 °C, respectively. The results confirm that this temperature range is suitable for biochar gasification.

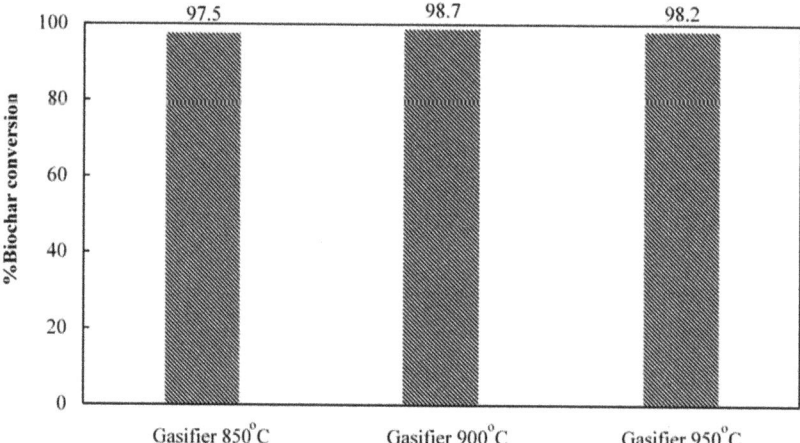

Figure 2. %Biochar conversion at different gasification temperatures using a $H_2O:O_2:C$ feed molar ratio of 0.25:0.25:1.

Product compositions obtained from the gasifier at different temperatures are shown in Figure 3. For the range of gasification temperature investigated in this work, two main products, CO and CO_2, are obtained. The obtained products could be due to the water–gas reaction (Equation (4)) and the partial oxidation reaction (Equation (5)).

Water gas reaction

$$C\,(s) + H_2O\,(g) \rightleftharpoons CO\,(g) + H_2\,(g) \qquad \Delta H = 131 \text{ kJ/kmol} \qquad (4)$$

Partial oxidation reaction

$$2C\ (s) + O_2\ (g) \rightleftharpoons 2CO\ (g) \qquad \Delta H = -221\ kJ/kmol \qquad (5)$$

Increasing gasification temperature from 850 °C to 950 °C shows insignificant effects on the production of H_2, whereas a gradual increase of CO production is observed with the reduction of CO_2. This phenomenon could be attributed to the result of a favorable Boudouard reaction (Equation (6)) [12,20]:

Boudouard reaction

$$C\ (s) + CO_2\ (g) \rightleftharpoons 2CO\ (g) \qquad \Delta H = 172\ kJ/kmol \qquad (6)$$

As seen from the above results, very small amounts of hydrogen can be obtained with solely biochar gasification. As a consequence, upgrading hydrogen production would further investigated by combining with steam reforming reaction. For gasification reaction, it was shown that almost complete conversion of biochar can be obtained in the range of gasification temperature investigated in this work, 850 °C to 950 °C. For optimistic reasons, a gasification temperature of 900 °C was chosen for investigating other effects on syngas production.

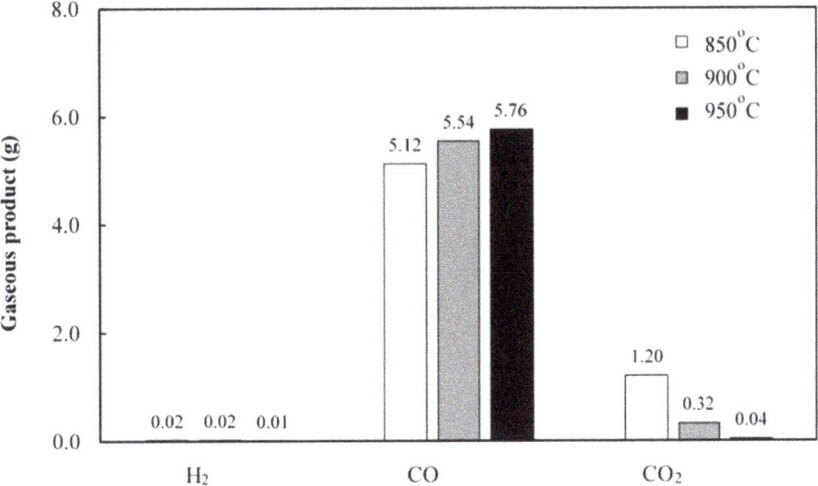

Figure 3. Gasification of biochar at different gasification temperatures ($H_2O:O_2:C$ feed molar ratio of 0.25:0.25:1 under atmospheric pressure).

3.2. Effect of Combined Gasification and Reforming Reaction

As shown in the previous section, biochar gasification can yield insignificant amount of H_2, to enhance the production of H_2, reforming reactor was introduced into the gasification system. In this investigation, to prove the concept of our simulation works [7] and to introduce process integration concept, three different packing patterns of sorbent and catalyst were studied, as shown in Figure 4. In Figure 4a, the catalyst and the sorbent were packed separately and the catalyst was packed on top of the sorbent, designated as the combined biomass gasifier and water-gas shift with Post-CO_2 recycle (CBGR-PostCO_2). In Figure 4b, the sorbent was placed on top of the catalyst, designated as the combined biomass gasifier and water-gas shift with Pre-CO_2 recycle (CBGR-PreCO_2), and in Figure 4c the developed one-body of combined catalyst with sorbent was introduced into the system, designated as the combined biomass gasifier and water-gas shift with multifunctional-CO_2 recycle (CBGR-SimulCO_2). In order to utilize CO_2, in this section, CO_2 was also used as co-gasifying agent together with H_2O and O_2. In this work, performances of each combined system were investigated in terms of syngas production and CO_2 emission ratio at a fixed gasification temperature of 900 °C, reforming temperature of 600 °C, $H_2O:CO_2:O_2:C$ feed molar ratio of 0.5:0.5:0.125:1, and NiO content of 12.5 wt.%.

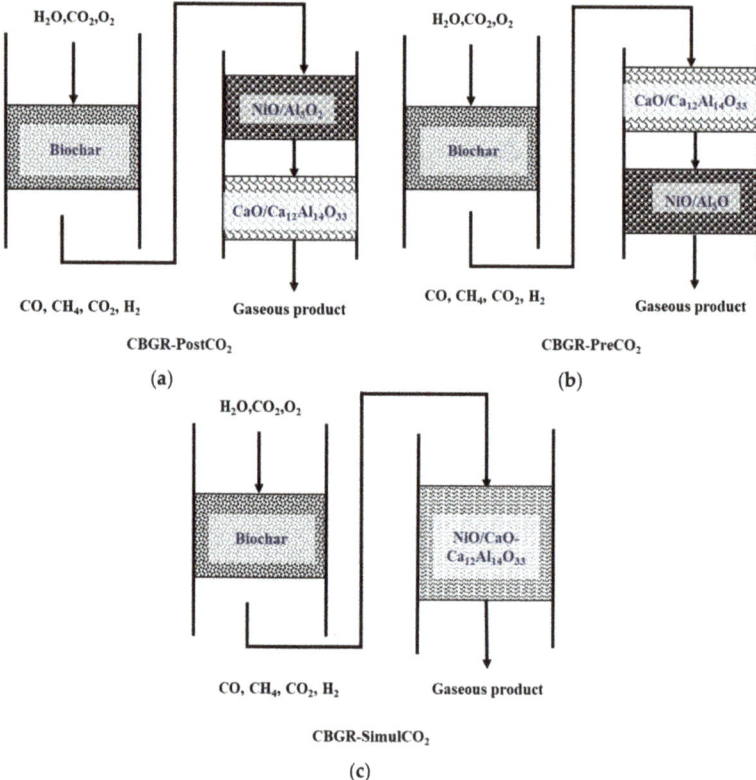

Figure 4. Patterns of sorbent and catalyst packing in the water–gas shift reactor (a) CBGR-PostCO$_2$, (b) CBGR-PreCO$_2$, and (c) CBGR-SimulCO$_2$.

Prior to running experiments, compositions of biochar were determined by proximate and ultimate analysis (Table 1). Compositions and surface textural properties of the synthetic materials were examined by XRD (Figure 5) and BET surface area analysis (Table 2), respectively. The results show XRD peaks corresponding to CaO at 2θ = 32.204, 37.347, and 64.154, $Ca_{12}Al_{14}O_{33}$ at 2θ = 18.052, 54.972, and 62.634, and NiO at 2θ = 37.249, 43.297, 62.934, and 67.271 [17,21]. Note that Ca(OH)$_2$ peaks, which are assigned at 2θ = 28.672, 34.102, and 47.121, are observed in the XRD pattern due to the fact that CaO is a hygroscopic material. The BET surface area of 12.5 wt.% NiO/CaO-Ca$_{12}$Al$_{14}$O$_{33}$ is 13.5 m^2/g, that of CaO-Ca$_{12}$Al$_{14}$O$_{33}$ is 5.91 m^2/g, and that of 12.5 wt.% NiO/Al$_2$O$_3$ is 59.1 m^2/g, respectively.

Table 1. Proximate and ultimate analysis results of biochar.

Proximate (wt.%)		Ultimate (wt.%)	
Moisture	5.30	C	80.20
Volatile matters	36.26	H	2.83
Fixed carbon	56.40	O (balance)	16.39
Ash	2.05	N	0.58

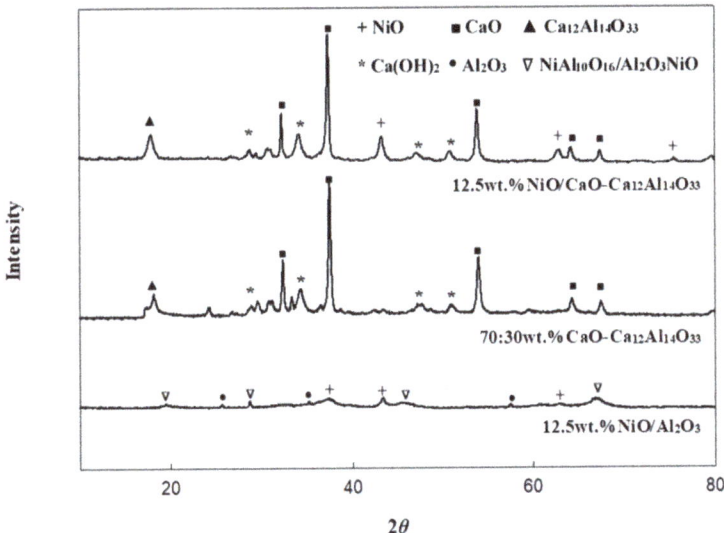

Figure 5. XRD patterns of NiO/Al$_2$O$_3$, CaO/Ca$_{12}$Al$_{14}$O$_{33}$, and 12.5 wt.% Ni/CaO-Ca$_{12}$Al$_{14}$O$_{33}$.

Table 2. Physical properties of materials from BET measurements.

Sample	Surface Area (m^2/g)	Pore Volume (cm^3/g)	Pore Size (nm)
12.5 wt.% NiO/Al$_2$O$_3$	59.1	0.150	0.09
CaO-Ca$_{12}$Al$_{14}$O$_{33}$	5.91	0.009	0.13
12.5 wt.% NiO/CaO-Ca$_{12}$Al$_{14}$O$_{33}$	13.5	0.016	0.16

As shown in Figure 6, the addition of reforming system (regardless of combination pattern) can provide higher H$_2$ production when compared with solely gasification reaction shown in Section 3.1. This could be due to the result of water–gas shift reaction (Equation (7)), where the main gasification product, CO, is further reacted with steam to form H$_2$ and CO$_2$ in the steam reforming reactor.

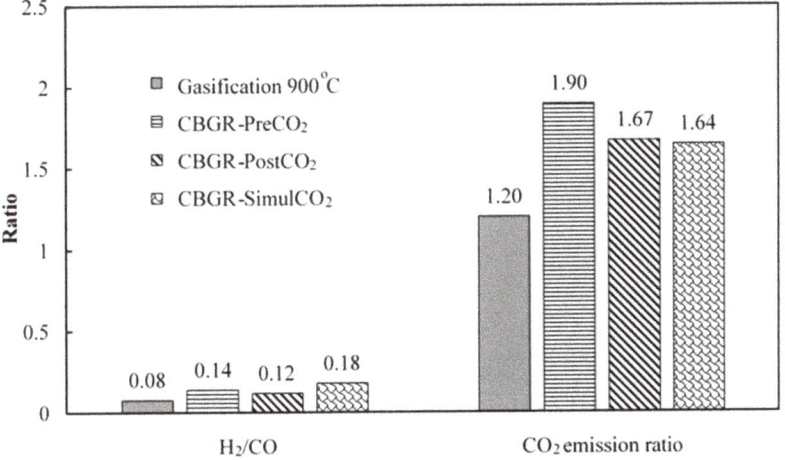

Figure 6. H$_2$/CO and CO$_2$ emission ratio from different proposed systems (H$_2$O:CO$_2$:O$_2$:C feed molar ratio of 0.5:0.5:0.125:1, gasification at 900 °C, reforming at 600 °C and ambient pressure).

Water–gas shift reaction

$$CO\ (g) + H_2O\ (g) \rightleftharpoons CO_2\ (g) + H_2\ (g) \qquad \Delta H = -41\ \text{kJ/kmol} \qquad (7)$$

Combining gasification with reforming system in different patterns shows the effect on syngas production and CO_2 emission ratio as demonstrated in Figure 6. The CBGR-PreCO$_2$ offers higher H_2 than the CBGR-PostCO$_2$, which is in good agreement with our simulation results proposed previously [7]. The enhancement of H_2 is believed to be mainly due to the result of the water–gas shift reaction. For the CBGR-PreCO$_2$ system, CO_2 was removed from the system prior to the water–gas shift reaction, leading to a favorable forward water–gas shift reaction. On the other hand, the CO_2 emission from CBGR-PreCO$_2$ is higher than CBGR-PostCO$_2$. This could be because the produced CO_2 is partly adsorbed by CaO-based sorbent in the CBGR-PostCO$_2$ system. Overall, the CBGR-SimulCO$_2$ system offers the highest H_2 production when compared with the CBGR-PreCO$_2$ system and the CBGR-PostCO$_2$ systems. This observation is due to the effect of the sorption-enhanced water–gas shift reaction; simultaneous removal of CO_2 can overcome the limitation of water–gas shift reaction (SEWGS) by inducing the system to proceed forward according to Le Chatelier's principle. More interestingly, the CO_2 emission ratio of the CBGR-SimulCO$_2$ system is found to be minimal, which could be attributed to greater CO_2 sorption capacity as mass transfer is favorable in the case of using one-body multi-functional material.

CO_2 adsorption

$$CaO\ (s) + CO_2\ (g) \rightleftharpoons CaCO_3\ (s) \qquad \Delta H = -178.2\ \text{kJ/kmol} \qquad (8)$$

As seen above, applying sorption-enhanced reaction (CBGR-SimulCO$_2$) system by introducing one-body multi-functional material can slightly increase H_2/CO with the reduction of the CO_2 emission from the system when compared with other sorption systems. However, all patterns provide H_2/CO ratios less than 0.18. This might be due to this biochar (H content is 2.83 wt.% from the ultimate analysis result, Table 1) not being favorable as feedstock for the production of syngas containing high hydrogen content. Nevertheless, the effect of operating conditions, including amount of catalyst, sorption-enhanced reaction temperature, and feed ratio of gasifying agent, were investigated for the combined gasification with SEWGS system.

3.3. Effect of Catalyst Amount

In this section, the effect of amount of catalyst on gaseous production, syngas H_2/CO ratio, and CO_2 emission ratio was studied. Figure 7 shows compositions of gaseous product for different wt.% of NiO. Comparative amounts of hydrogen production are obtained for all NiO contents, whereas the maximum of CO production is found with 12.5 wt.% Ni/CaO-Ca$_{12}$Al$_{14}$O$_{33}$. This result could be because the 12.5 wt.% Ni/CaO-Ca$_{12}$Al$_{14}$O$_{33}$ possesses the highest BET surface area, resulting in higher active surface exposure, as shown in Table 3. Large amount of NiO (17.5 wt.%) could block the small pores of the support, leading to the reduction of surface area as well as pore volume with an increase of average pore size diameter. For the 7.5 wt.% Ni/CaO-Ca$_{12}$Al$_{14}$O$_{33}$, small amounts of Ni cannot help prevent the agglomeration of CaO particles, resulting in lower surface area (Table 3) and dense packing particles (Figure 8).

Table 3. Physical properties of multi-functional materials for different NiO contents.

Sample	Surface Area (m^2/g)	Pore Volume (cm^3/g)	Pore Size (nm)
7.5 wt.% NiO/CaO-Ca$_{12}$Al$_{14}$O$_{33}$	11.70	0.026	0.11
12.5 wt.% NiO/CaO-Ca$_{12}$Al$_{14}$O$_{33}$	13.50	0.016	0.16
17.5 wt.% NiO/CaO-Ca$_{12}$Al$_{14}$O$_{33}$	12.45	0.023	0.11

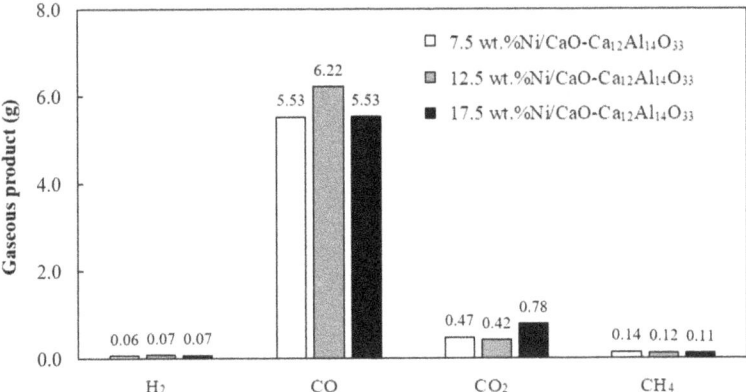

Figure 7. Product composition at different wt.% of Ni on Ni/CaO-Ca$_{12}$Al$_{14}$O$_{33}$ using a H$_2$O:CO$_2$:O$_2$:C feed molar ratio of 0.5:0.1:0.125:1, gasification at 900 °C, and sorption-enhanced water–gas shift reaction (SEWGS) at 600 °C.

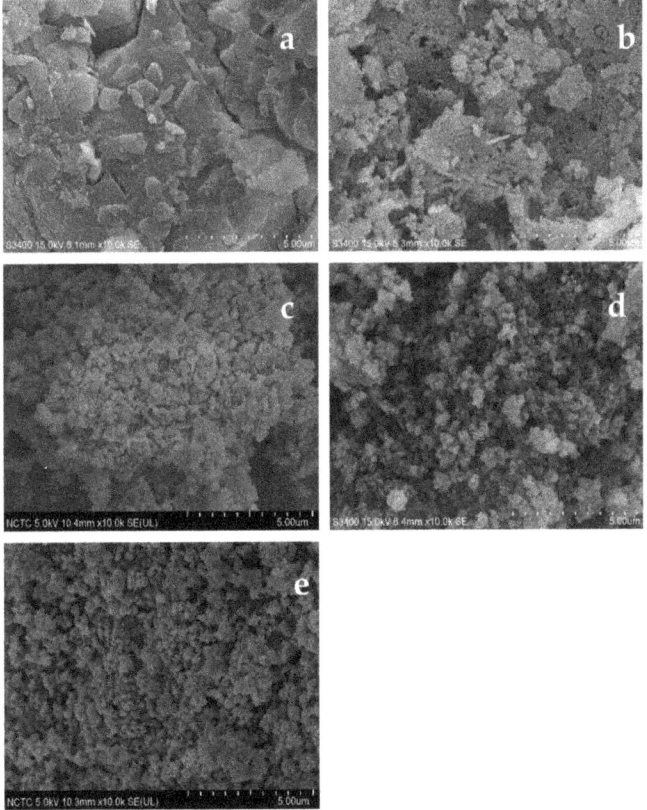

Figure 8. SEM images of fresh sample materials; (**a**) Ni/Al$_2$O$_3$, (**b**) CaO-Ca$_{12}$Al$_{14}$O$_{33}$, (**c**) 7.5 wt.% Ni/CaO-Ca$_{12}$Al$_{14}$O$_{33}$, (**d**) 12.5 wt.% Ni/CaO-Ca$_{12}$Al$_{14}$O$_{33}$, and (**e**) 17.5 wt.% Ni/CaO-Ca$_{12}$Al$_{14}$O$_{33}$.

Figure 9 shows CO$_2$ emission ratio of different NiO contents. The results show that the 12.5 wt.% Ni/CaO-Ca$_{12}$Al$_{14}$O$_{33}$ provides minimum CO$_2$ emission ratio, which could be due to the result of high performance of SEWGS reaction. Lower CO_2 adsorption observed with the 7.5 wt.% Ni/CaO-$Ca_{12}Al_{14}O_{33}$ and the 17.5 wt.% Ni/CaO-$Ca_{12}Al_{14}O_{33}$ could be attributed to lower surface area as shown in Table 3.

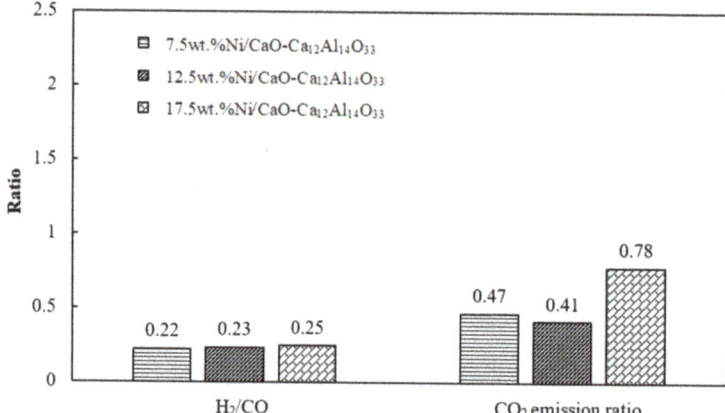

Figure 9. H_2/CO and CO_2 emission ratios at different wt.% of Ni of Ni/CaO-$Ca_{12}Al_{14}O_{33}$ using a H_2O:CO_2:O_2:C feed molar ratio of 0.5:0.1:0.125:1, gasification at 900 °C, and SEWGS at 600 °C.

3.4. Effect of Sorption-Enhanced Water–Gas Shift (SEWGS) Temperature

As seen from the previous section, combining gasification with SEWGS reaction with the use of 12.5 wt.% Ni/CaO-$Ca_{12}Al_{14}O_{33}$ can provide greater H_2/CO ratio with lower CO_2 emission. In this section, the effect of SEWGS temperature on H_2/CO ratio and CO_2 emission ratio was investigated. Figure 10 shows product compositions obtained at different SEWGS temperatures. Increasing SEWGS temperature from 500 °C to 650 °C does not affect the production of hydrogen or the quality of syngas, as comparative values are observed. The reduction of CO could be due to the reactions between CO and H_2O (Equation (7)) and CO and H_2 (reversed Equation (9)) which lead to the formation of CH_4. Increasing temperature results in the decrease of CH_4 due to exothermic reaction of Equation (9). The CO_2 emission is found to decrease with increasing SEWGS temperature from 500 to 650 °C (Figure 11). This observation could be due to the result of the suitable CO_2 sorption condition of CaO sorbent at a high temperature of 650 °C [22,23].

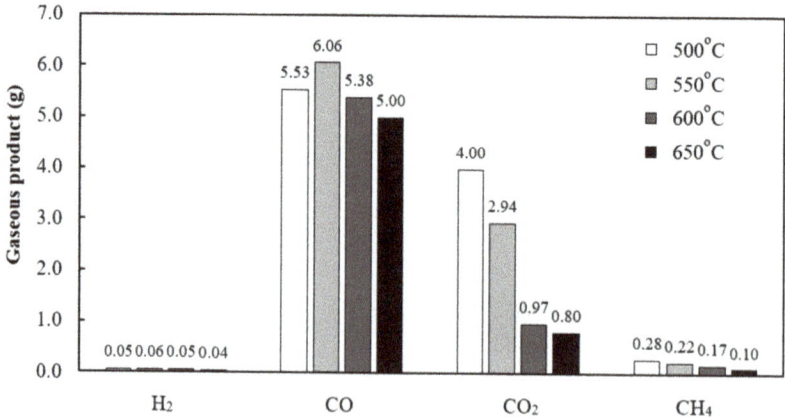

Figure 10. Product composition obtained from different SEWGS temperatures using a H_2O:CO_2:O_2:C feed molar ratio of 0.5:0.5:0.125:1 and gasification at 900 °C.

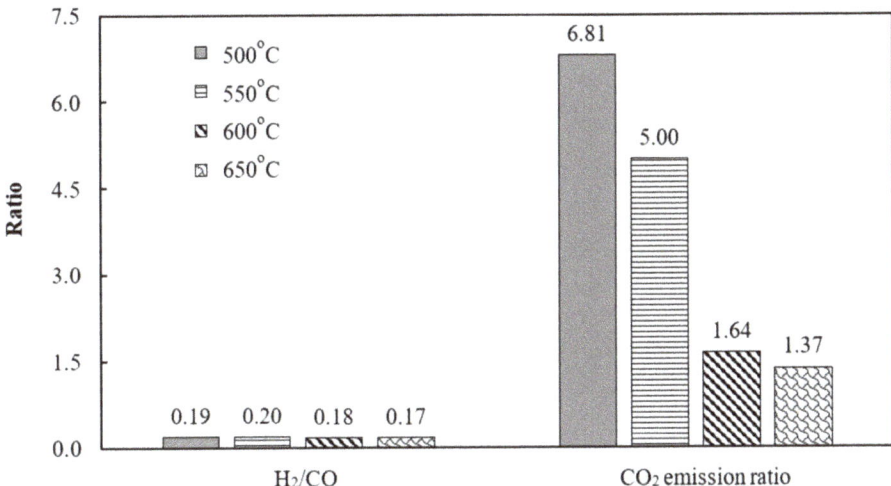

Figure 11. H_2/CO and CO_2 emission ratios at different SEWGS temperatures using a $H_2O:CO_2:O_2:C$ feed molar ratio of 0.5:0.5:0.125:1 and gasification at 900 °C.

CO_2 reforming

$$CH_4\ (g) + CO_2\ (g) \rightleftharpoons 2CO\ (g) + 2H_2\ (g) \qquad \Delta H = 247\ kJ/kmol \qquad (9)$$

3.5. Effect of Gasifying Agent

As the gasifying agent is one factor that can affect gasification of biomass [24], in this work, we investigated the effect of introducing CO_2 as co-gasifying agent in order to utilize the CO_2. The feed molar ratio of co-feed gasifying agent was fixed at $H_2O:O_2:C = 0.5:0.125:1$, while the CO_2/C molar ratio was varied between 0.1 and 0.5:1 using 12.5 wt.% Ni/CaO-Ca$_{12}$Al$_{14}$O$_{33}$ at a gasification temperature of 900 °C and SEWGS temperature of 600 °C (Section 3.5.1). And effect of H_2O feed as gasifying agent was investigated by varying the $H_2O:C$ ratio between 0.5–1.5:1 at a fixed $CO_2:O_2:C$ feed molar ratio of 0.1:0.125:1, gasification temperature of 900 °C, and SEWGS temperature of 600 °C (Section 3.5.2).

3.5.1. Effect of CO_2 Feed

Figure 12 presents product composition obtained from the reaction with different amounts of CO_2 feed. The results show that CO increases with increasing CO_2/C ratio parallel with an increase of CO_2 emission ratio (Figure 13). This observed result could be attributed to the reverse Boudouard reaction (Equation (1)). It is noted that although higher amounts of H_2 produced from the system could be obtained due to the result of water–gas shift reaction, negligible amounts of produced H_2 are still observed. This result might be because of insufficient steam feed into the system, leading to a smaller contribution of the water–gas shift reaction. Our observation is consistent with the results obtained from a thermodynamic study of lignite coal gasification reported by Kale et al. [25], where an increase of CO_2/C feed mole ratio from 0 to 1 led to a decrease of H_2/CO ratio from 3.04 to 0.7. It is also noted that CH_4 is observed in gaseous products, implying that reverse Boudouard reaction (Equation (1)) could occur due to the addition of CO_2, resulting in higher production of CO which could further react with the produced H_2 to form CH_4 and CO_2 (reversed Equation (9)).

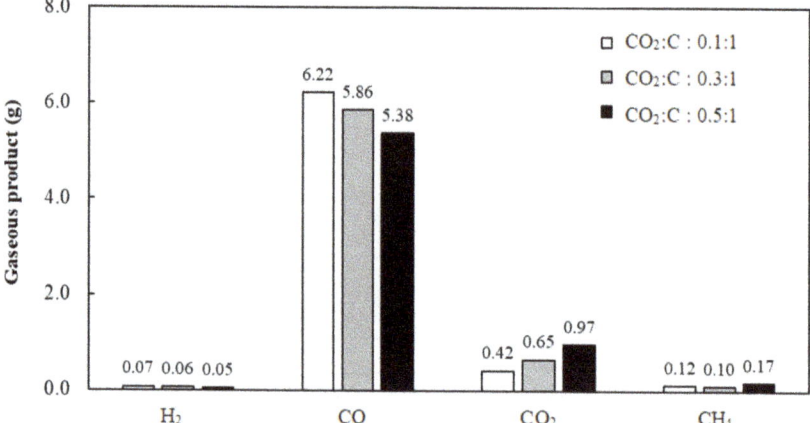

Figure 12. Product composition at different $CO_2:C$ ratios using a $H_2O:CO_2:O_2:C$ feed molar ratio of 0.5:0.1–0.5:0.125:1, gasification at 900 °C, and SEWGS at 600 °C.

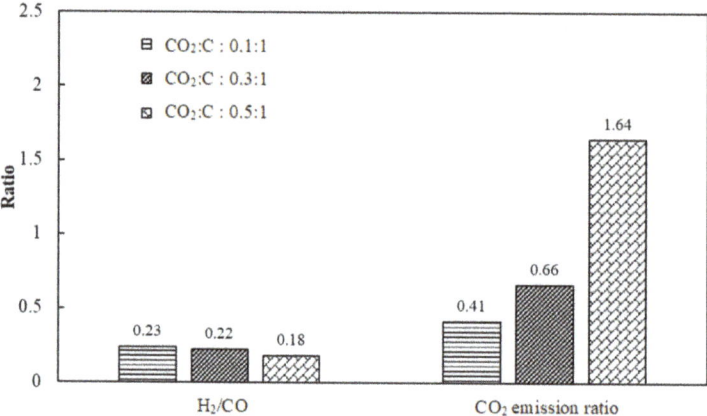

Figure 13. H_2/CO and CO_2 emission ratios at different $CO_2:C$ ratios using a $H_2O:CO_2:O_2:C$ feed molar ratio of 0.5:0.1–0.5:0.125:1, gasification at 900 °C, and SEWGS at 600 °C.

3.5.2. Effect of H_2O Feed

In Figure 14, increasing of H_2O feed is expected to enhance the production of H_2 due to a water gas reaction (Equation (4)) and water–gas shift reaction (Equation (7)); however, insignificant H_2 production is observed. This result might be due to insufficient H_2O feed as discussed previously. Nevertheless, quality of syngas (H_2/CO ratio) is found to increase with increasing H_2O feed (Figure 15). This observation could possibly be due to the produced CO reacting with the produced H_2 via the reversed CO_2 reforming reaction (reversed Equation (9)) as evidenced by the reduction of CO and the increase of CO_2 shown in Figure 14.

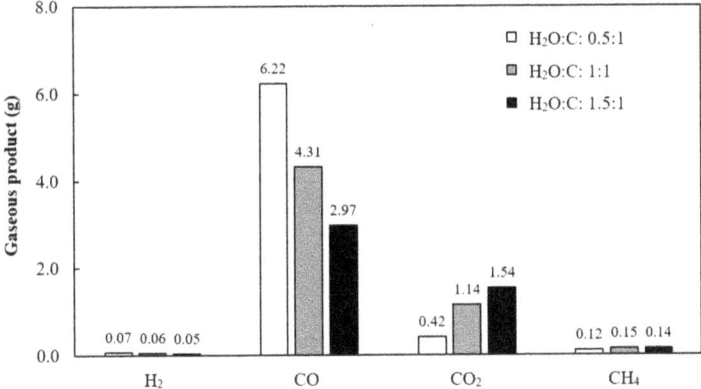

Figure 14. Product composition at different $H_2O:C$ ratios using a $H_2O:CO_2:O_2:C$ feed molar ratio of 0.5–1.5:0.1:0.125:1, gasification at 900 °C and SEWGS at 600 °C.

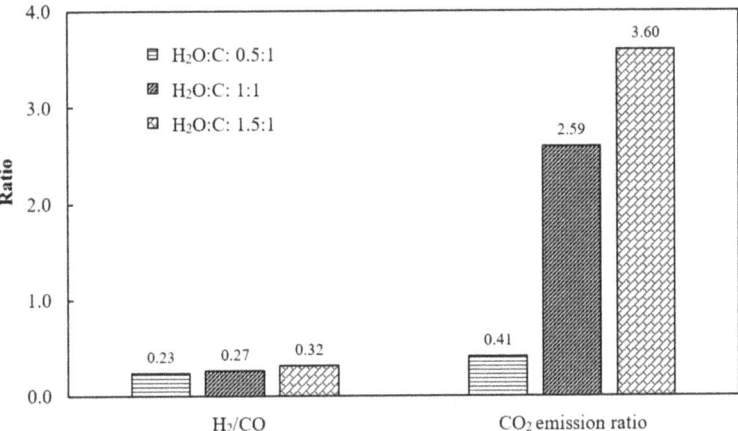

Figure 15. H_2/CO and CO_2 emission ratios at different $H_2O:C$ ratios using a $H_2O:CO_2:O_2:C$ feed molar ratio of 0.5–1.5:0.1:0.125:1, gasification at 900 °C, and SEWGS at 600 °C.

4. Conclusions

Our studies investigated the production of syngas from biochar using the combined gasification and sorption-enhanced water–gas shift reaction. The optimum gasification temperature was 900 °C with a $H_2O:O_2:C$ feed molar ratio of 0.25:0.25:1. The one-body material combining catalyst with sorbent (CBGR-SimulCO$_2$) could provide the highest H_2/CO ratio as well as the lowest CO_2 emissions when compared to the other systems. In addition, the effect of sorption-enhanced water–gas shift temperature was shown to affect CO_2 emissions. Increasing the operating temperature from 500 to 650 °C led to a decrease of the CO_2 emission ratio. Increasing the CO_2/C ratio from 0.1 to 0.5 resulted in an increase of CO production with a lower CO_2 emission ratio. In addition, increasing the H_2O/C ratio from 0.5 to 1.5 provided higher syngas production, with H_2/CO ratios of 0.23 and 0.32, respectively.

Author Contributions: Conceptualization, Supawat Vivanpatarakij; Data curation, Supanida Chimpae, Suwimol Wongsakulphasatch, Supawat Vivanpatarakij and Fasai Wiwatwongwana; Formal analysis, Suwimol Wongsakulphasatch, Supawat Vivanpatarakij, Thongchai Glinrun and Weerakanya Maneeprakorn; Funding acquisition, Suttichai Assabumrungrat; Investigation, Supanida Chimpae and Thongchai Glinrun; Methodology, Supanida Chimpae, Suwimol Wongsakulphasatch and Supawat Vivanpatarakij; Project administration, Suttichai Assabumrungrat; Resources, Thongchai Glinrun, Fasai Wiwatwongwana and

Weerakanya Maneeprakorn; Supervision, Suwimol Wongsakulphasatch, Supawat Vivanpatarakij and Suttichai Assabumrungrat; Validation, Thongchai Glinrun and Weerakanya Maneeprakorn; Writing—original draft, Supanida Chimpae, Suwimol Wongsakulphasatch, Supawat Vivanpatarakij and Suttichai Assabumrungrat.

Funding: The authors would like to acknowledge funding support from the Ratchadapisek Sompoch Endowment Fund 2016 of Chulalongkorn University (CU-59-003-IC), and King Mongkut's University of Technology North Bangkok (contract no. KMUTNB-KNOW-61-029). S.W. and S.A. also wish to acknowledge the "Research Chair Grant" of the National Science and Technology Development Agency (NSTDA).

Conflicts of Interest: The authors declare no conflict of interest.

References

1. Rodrigues, M.; Faaij, A.P.C.; Walter, A. Technol-economic analysis of co-fired biomass integrated gasification/combined cycle systems with inclusion of economies of scale. *Energy* **2003**, *28*, 1229–1258. [CrossRef]
2. Hamelinck, C.N.; Faaij, A.P.C.; den Uil, H.; Boerrigter, H. Production of FT transportation fuels from biomass; technical options, process analysis and optimization, and development potential. *Energy* **2004**, *29*, 1743–1771. [CrossRef]
3. Adams, T.A.; Hoseinzade, L.; Madabhushi, P.B.; Okeke, I.J. Comparison of CO_2 capture approaches for fossil-based power generation: Review and Meta-study. *Processes* **2017**, *5*, 44. [CrossRef]
4. Acharya, B.; Dutta, A.; Basu, P. Chemical-looping gasification of biomass for hydrogen-enriched gas production with in-process carbon dioxide capture. *Energy Fuels* **2009**, *23*, 5077–5083. [CrossRef]
5. Pang, Y.; Hammer, T.; Müller, D.; Karl, J. Investigation of nonthermal plasma assisted charcoal gasification for production of hydrogen-rich syngas. *Processes* **2019**, *7*, 114. [CrossRef]
6. Chaiwatanodom, P.; Vivanpatarakij, S.; Assabumrungrat, S. Thermodynamic analysis of biomass gasification with CO_2 recycle for synthesis gas production. *Appl. Energy* **2014**, *14*, 10–17. [CrossRef]
7. Kraisornkachit, P.; Vivanpatarakit, S.; Amornraksa, S.; Simasatitkul, L.; Assabumrungrat, S. Performance evaluation of different combined systems of biochar gasifier, reformer and CO_2 capture unit for synthesis gas production. *Int. J. Hydrog. Energy* **2016**, *41*, 13408–13418. [CrossRef]
8. Ma, Z.; Zhang, S.; Xie, D.; Yan, Y. A novel integrated process for hydrogen production from biomass. *Int. J. Hydrog. Energy* **2014**, *39*, 1274–1279. [CrossRef]
9. Waheed, Q.M.K.; Williams, P.T. Hydrogen production from high temperature pyrolysis/steam reforming of waste biomass: Rice husk, sugar cane bagasse, and wheat straw. *Energy Fuels* **2013**, *27*, 6695–6704. [CrossRef]
10. Gao, N.; Lia, A.; Quan, C.; Gao, F. Hydrogen-rich gas production from biomass steam gasification in an updraft fixed-bed gasifier combined with a porous ceramic reformer. *Int. J. Hydrog. Energy* **2008**, *33*, 5430–5438. [CrossRef]
11. Liu, S.; Zhu, J.; Chen, M.; Xin, W.; Yang, Z.; Kong, L. Hydrogen production via catalytic pyrolysis of biomass in a two-stage fixed bed reactor system. *Int. J. Hydrog. Energy* **2014**, *39*, 13128–13135. [CrossRef]
12. Waheed, Q.M.K.; Wu, C.; Williams, P.T. Hydrogen production from high temperature steam catalytic gasification of bio-char. *J. Energy Inst.* **2016**, *89*, 222–230. [CrossRef]
13. Song, C.; Liu, Q.; Ji, N.; Deng, S.; Zhao, J.; Li, Y.; Song, Y. Alternative pathways for efficient CO_2 capture by hybrid processes—A review. *Renew. Sustain. Energy Rev.* **2018**, *82*, 215–231. [CrossRef]
14. La Villetta, M.; Costa, M.; Massarotti, N. Modelling approaches to biomass gasification: A review with emphasis on the stoichiometric method. *Renew. Sustain. Energy Rev.* **2017**, *74*, 71–88. [CrossRef]
15. Indrawan, N.; Thapa, S.; Bhoi, P.R.; Huhnke, R.L.; Kumar, A. Engine power generation and emission performance of syngas generated from low-density biomass. *Energy Convers. Manag.* **2017**, *148*, 593–603. [CrossRef]
16. Ruiz, J.A.; Juárez, M.C.; Morales, M.P.; Muñoz, P.; Mendívil, M.A. Biomass gasification for electricity generation: Review of current technology barriers. *Renew. Sustain. Energy Rev.* **2013**, *18*, 174–183. [CrossRef]
17. Guohui, L.; Linjie, H.; Josephine, M.H. Comparison of reducibility and stability of alumina-supported Ni catalysts prepared by impregnation and co-precipitation. *Appl. Catal. A* **2006**, *301*, 16–24.
18. Cong, L.; Ying, Z.; Ning, D.; Chuguang, Z. Enhanced cyclic stability of CO_2 adsorption capacity of CaO-based sorbents using La_2O_3 or $Ca_{12}Al_{14}O_{33}$ as additives. *Korean J. Chem. Eng.* **2011**, *28*, 1042–1046.
19. Changjun, Z.; Zhiming, Z.; Cheng, Z.; Fang, X. Sol-gel-derived, $CaZrO_3$-stabilized Ni/CaO-$CaZrO_3$ bifunctional catalyst for sorption-enhanced steam methane reforming. *Appl. Catal. B* **2016**, *196*, 16–26.

20. Howaniec, N.; Smolinski, A.; Stanczyk, K.; Pichlak, M. Steam co-gasification of coal and biomass derived chars with synergy effect as an innovative way of hydrogen-rich gas production. *Int. J. Hydrog. Energy* **2011**, *36*, 14455–14463. [CrossRef]
21. Pan, X.; Miaomiao, X.; Zhenmin, C.; Zhiming, Z. CO_2 capture performance of CaO-based sorbents prepared by a pol–gel method. *Ind. Eng. Chem. Res.* **2013**, *52*, 12161–12169.
22. Lu, H.; Reddy, E.P.; Smirniotis, P.G. Calcium oxide based sorbents for capture of carbon dioxide at high temperatures. *Ind. Eng. Chem. Res.* **2006**, *45*, 3944–3949. [CrossRef]
23. Yu, C.H.; Huang, C.H.; Tan, C.S. A review of CO_2 capture by absorption and adsorption. *Aerosol Air Qual. Res.* **2012**, *12*, 745–769. [CrossRef]
24. Butterman, H.C.; Castaldi, M.J. CO_2 as a carbon neutral fuel source via enhanced biomass gasification. *Environ. Sci. Technol.* **2009**, *43*, 9030–9037. [CrossRef] [PubMed]
25. Kale, G.R.; Kulkarni, B.D.; Chavan, R.N. Combined gasification of lignite coal: Thermodynamic and application study. *J. Taiwan Inst. Chem. Eng.* **2013**, *45*, 163–173. [CrossRef]

© 2019 by the authors. Licensee MDPI, Basel, Switzerland. This article is an open access article distributed under the terms and conditions of the Creative Commons Attribution (CC BY) license (http://creativecommons.org/licenses/by/4.0/).

Article

Conceptual Design of Pyrolytic Oil Upgrading Process Enhanced by Membrane-Integrated Hydrogen Production System

Bo Chen [1,*], Tao Yang [1], Wu Xiao [2] and Aazad khan Nizamani [2]

1   R & D Center of Hydroprocesing technology, SINOPEC Dalian Research Institute of Petroleum and Petrochemicals, Dalian 116045, China; yt.fshy@sinopec.com
2   Chemical Engineering Department, Dalian University of Technology, Dalian 116045, China; wuxiao@dlut.edu.cn (W.X.); aazadniz@mail.dlut.edu.cn (A.k.N.)
*   Correspondence: chenbo.dshy@sinopec.com; Tel.: +86-411-3969-9517

Received: 18 April 2019; Accepted: 10 May 2019; Published: 14 May 2019

Abstract: Hydrotreatment is an efficient method for pyrolytic oil upgrading; however, the trade-off between the operational cost on hydrogen consumption and process profit remains the major challenge for the process designs. In this study, an integrated process of steam methane reforming and pyrolytic oil hydrotreating with gas separation system was proposed conceptually. The integrated process utilized steam methane reformer to produce raw syngas without further water–gas-shifting; with the aid of a membrane unit, the hydrogen concentration in the syngas was adjusted, which substituted the water–gas-shift reactor and improved the performance of hydrotreater on both conversion and hydrogen consumption. A simulation framework for unit operations was developed for process designs through which the dissipated flow in the packed-bed reactor, along with membrane gas separation unit were modeled and calculated in the commercial process simulator. The evaluation results showed that, the proposed process could achieve 63.7% conversion with 2.0 wt% hydrogen consumption; the evaluations of economics showed that the proposed process could achieve 70% higher net profit compared to the conventional plant, indicating the potentials of the integrated pyrolytic oil upgrading process.

Keywords: hydrogen production; pyrolytic oil hydro-processing; process modeling; syngas

1. Introduction

The growing demand of social development and energy consumption raises the request for energies. Although fossil fuel remains the world's primary energy source, the CO_2 emission during its combustion process has been boosting the global warming effect. Therefore, biomass, as an alternative renewable energy source, has been considered as a potential solution for energy supply, all attributed to its advantages on low carbon footprint and closed-loop carbon cycles [1].

Biomass is usually liquified to pyrolytic oil for further treatment and utilization. Pyrolytic oil could be produced from various sources [2], such as wood waste, energy crops or other organic materials. The production process of pyrolytic oil is heating biomass under anaerobic condition with temperatures above 500 °C [3]; the liquefied oil from that process normally contains oxygen, resulting in non-volatility, corrosiveness, immiscibility and thermal instability [4–6]. Therefore, upgrading processes for pyrolytic oil are required for environmental-friendly utilization.

Hydrotreatment is one of the most efficient methods to modify the molecular structure of pyrolytic oil [7,8]. Through hydro-processing, impurities such as sulfur, oxygen and nitrogen could be removed [9]; proper hydrotreatment could also make pyrolytic oil lighter by converting heavy components, such as tars and heavy non-volatiles, into lighter oil cuts, to raise the quality of oils and enable them to be utilized as chemical materials [10]. Hydrogen is one of the key reactants in the hydrotreatment process, most of which was produced through reforming in a plant [11]. Steam methane reforming (SMR) is the most widely used hydrogen production method due to the environmental-friendly production process. Two reaction steps are required by the SMR process, which are reforming and water–gas-shift (WGS); a gas separation system followed the latter reaction step to store CO_2 and purify hydrogen through pressure swing adsorption (PSA) [12]. The conventional process schematic was shown in Figure 1a. Although PSA could produce hydrogen with >99mol% purity, the high operational cost of it limits the application [13]. Conventional hydro-processing reactors require a hydrogen stream with 80mol%–95mol% purity to sustain the upgrading process of pyrolytic oil, and thus the PSA separation unit may not be adequate for such process. In addition, the water, CO and CO_2 are generated in pyrolytic oil upgrading reactions as byproducts, which could contribute to hydrogen production through an in-situ WGS reaction; therefore it is feasible to employ syngas (a raw product output from SMR reactor) as a hydrogen supplier for the pyrolytic oil hydrotreating (HT) process.

The application of syngas HT was investigated by Fu et al. [14] and the results showed promising potentials. They employed syngas to hydrotreat the liquefied coal, and the conversion was similar to pure hydrogen; they suggested that the water content showed significant impacts on the performance, which could reduce hydrogen consumption by an in-situ WGS reaction. However, there are still several issues that await further solutions for the process. Biomass-derived pyrolytic oil has various properties over a long time duration [15], and therefore the syngas composition may not suit all HT processes; the cost-sensitive nature of pyrolytic oil requires high efficiency on production and low capital expense (CAPEX) and operation expense (OPEX) to compensate the cost on feedstock [16]. Therefore, intensive research efforts on new process designs and optimizations are required to promote the application of the pyrolytic oil HT process [17]. Hydrogen takes about 90% of the total OPEX, the production and recovery of which determine the profit of whole process; therefore, the hydrogen production and recovery system deserve a thorough optimization for the pyrolytic oil upgrading process. The economic issues would be the primary concern for the HT process, and employing syngas as the hydrogen source could be an alternative solution. Nonetheless, the feasibility of hydrotreatment with syngas enables it to consider omitting the WGS reactor to simplify the process design, but the demands of the plant on high-purity hydrogen (>99mol%) accompanying it would be an obstruction for the application.

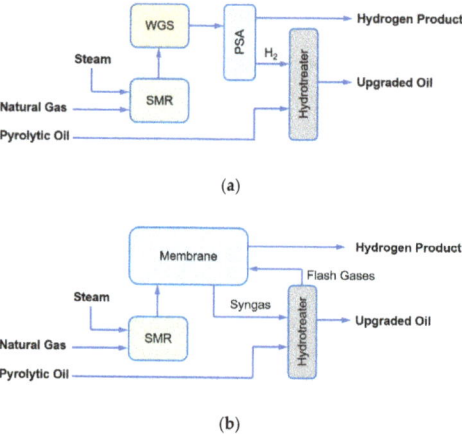

Figure 1. Design and schematics of the integrated hydrogen production and pyrolytic oil upgrading process. (**a**) Conventional process with a water–gas-shift (WGS) reactor and pressure swing adsorption (PSA) unit; (**b**) integrated steam methane reforming-hydrotreating process without a WGS reactor.

In this study, an integrated steam methane reforming-hydrotreating (SMR-HT) pyrolytic oil upgrading process enhanced by membrane gas separation system was proposed and optimized to improve the upgrading efficiency, and the design schematic of the process is shown in Figure 1b. The syngas was produced by SMR reactor and purified by membrane gas separation unit subsequently; the sweetened syngas was applied to hydrotreat the pyrolytic oil feedstock in a packed-bed reactor, in which the oil was upgraded with presence of an in-situ WGS reaction; the flash gases were further recovered and purified to produce high-purity hydrogen, which improved the process efficiency by mutual production of pure hydrogen and upgraded oil. An axial-dispersion model with five-lump reaction kinetics and a WGS reaction was proposed to model the packed-bed pyrolytic oil HT reactor; through which the process was simulated and optimized in the commercial simulator. This study provided an alternative process design for the pyrolytic oil upgrading process; through the model-based process optimizations, the proposed theoretical framework could also provide guidance on future applications of pyrolytic oil upgrading processes.

2. Theory

2.1. Reaction Kinetics

The upgrading process of pyrolytic oil is complicated due to the complex compositions, and therefore lumping strategy is the optimal method to model such process. This study employed the reaction network [18] that was proposed and investigated by Stowe and Raal et al. [19,20]. The pyrolytic oil was cut into five lumps, and the properties are shown in Table 1. The Ni–Mo@Al$_2$O$_3$ catalyst was employed in the simulation for the HT reaction; the conversion network was presented in Figure 2 (in which heavy non-volatile is denoted HNV, light non-volatile is denoted LNV). The reaction rate constant was calculated with the Arrhenius equation as shown in Equation (1); the constants and parameters of the equations are shown in Table 2. The studied range of reaction temperature was 350–420 °C, and operating pressure was 3–10 MPa [21].

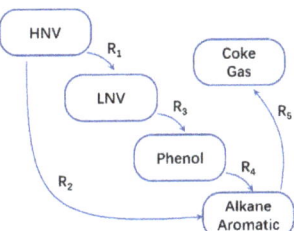

Figure 2. Reaction networks for pyrolytic oil upgrading.

Table 1. Properties of pyrolytic oil.

Item	Value
Mass Density, kg/m^3	779.8
Viscosity, cP	5.454
Composition, %	
H$_2$O	0.13
Gases	0.01
Aromatics	4.20
Phenol	30.53
LNV	44.58
HNV	20.55

Table 2. Parameters for pyrolytic upgrading reaction kinetics [18].

Reaction	Pre-Exponential Factor	Activation Energy, J/mol
R_1	8.8×10^3	7.5×10^4
R_2	6.5×10^5	8.5×10^4
R_3	3.1×10^5	9.0×10^4
R_4	1.9×10^3	6.8×10^4
R_5	1.6×10^4	7.5×10^4

The pyrolytic oil contains oxygen, which implies that a considerable amount of water would be generated by HT reactions during processing. At the operating temperature, water and syngas would induce an in-situ WGS reaction, which would produce hydrogen and CO_2. The reaction formula is

$$CO + H_2O \leftrightarrow CO_2 + H_2.$$

The additional hydrogen could promote the HT reactions, and hence improve the upgrading process.

The syngas was produced by methane reforming reaction. In this study, the SMR process was employed and modified for hydrogen production. The methane and steam were fed into the reforming reactor at a temperature of 600–620 °C, and the reaction formula is

$$CH_4 + H_2O \leftrightarrow CO + 3H_2.$$

Table 3. Parameters for reaction kinetics.

Reaction	Pre-Exponential Factor	Activation Energy, J/mol	Reverse Reaction Parameters [a] $A'/B'/C'/D'$
Steam methane reforming	6.0×10^2	3.3×10^4	−21/−23000/7.2/−0.0029
Water-gas-shift	5.5×10^2	4.2×10^4	−12/−5300/1/−0.0001

[a] Effects of the reverse reaction was calculated by equilibrium constant.

The rate constants of the SMR and WGS reactions are calculated by the Arrhenius equation

$$k = A \cdot \exp\left(-\frac{E_a}{RT}\right) \quad (1)$$

where, k is the reaction rate constant; A is the pre-exponential parameter; E_a is the activation energy; R is the gas constant; T is the temperature.

The reaction rate of the reaction is

$$r = k(f - f'/k') \quad (2)$$

where, f and f' is the concentration of the component; k' is the rate constant of the reverse reaction. The equilibrium constant of the forward and reverse reactions of SMR and WGS are governed by Equations (3) and (4)

$$\ln(K_{eq}) = A' + \frac{B'}{T} + C' \cdot \ln(T) + D' \cdot T \quad (3)$$

$$K_{eq} = k/k' \quad (4)$$

where the coefficients A', B', C' and D' are shown in Table 3.

The reforming reaction was endothermic, and a furnace was required for the reactor to maintain heat balance. The fuel gas to the furnace was taken from feed methane; in the proposed process, the hydrogen-rich gases were also considered as fuel gas to supply energy cost of the reactors.

2.2. Modeling of Reactors

The reforming reactor for natural gas is a tube reactor, which could be modeled by the inbuilt plug flow reactor (PFR) module in Aspen HYSYS (Aspen V10.0, Bedford, MA, USA). The reaction of steam reforming is endothermic, and the reaction heat was supplied by the furnace. The packed-bed reactor for pyrolytic oil HT was more complex due to the non-uniform flow distribution in the porous catalyst bed. The flow of gas and liquid through the catalyst bed was in a dissipated regime. In this study, the axial-dispersion model was employed to simulate the dissipated flow in the packed-bed reactor. The mass transfer in the reactor is governed by

$$\frac{\partial C}{\partial t} = D_e \frac{\partial^2 C}{\partial z^2} - u \frac{\partial C}{\partial z} + R \tag{5}$$

where, D_e is the axial-dispersion coefficient; C is the concentration of the component; z is the axial position; u is the fluid velocity; R is the source term. By integrating the reaction network that was described in Section 2.1, the HT process of the pyrolytic oil could be calculated in the packed-bed reactor. Equation (5) is a set of partial differential equations (PDEs); in this study, the PDEs were solved by FiPy, a finite volume method (FVM) toolbox developed by NIST (National Institute of Standards and Technology) in Python [22].

The reaction rates of the HT process are governed by the Arrhenius equation (Equation (1)). Accordingly, the source term could be expressed

$$r = \sum kC_i C_{H2}. \tag{6}$$

The boundary conditions to the PDEs are

$$\begin{aligned} -D_e \tfrac{dC}{dz} &= u(C - C_0),\ z = 0 \\ \tfrac{dC}{dz} &= 0,\ z = L. \end{aligned} \tag{7}$$

2.3. Modeling of Membrane Gas Separation

The materials of gas separation membrane are polymeric, and the mass transfer mechanism is governed by solution-diffusion. In such a scheme, the gas molecule is separated in the membrane matrix by selective permeation. The solution-diffusion mechanism is

$$J = D \cdot S \tag{8}$$

where J is the permeance; D is the diffusivity; S is the solubility.

The gas separation membrane is housed in a module, and the mass transfer equation in the module is

$$dN = J \cdot \Delta P \cdot ds \tag{9}$$

where, N is the permeation flux; P is the pressure difference; s is the membrane area. The pressure difference is calculated by partial pressure of components. A plug-flow assumption was applied by the model in the membrane lumen, and the pressure drop was calculated by the Hagen–Poiseuille equation.

The upwind finite difference method was applied to solve Equation (9). The validation of the models was investigated and discussed in previous works [23–27].

In this study, the hydrogen-selective membrane was employed for hydrogen enrichment. Polyimide was selected as the membrane material, and the gas permeation properties are shown in Table 4.

Table 4. Permeances of the gas separation membrane.

Component	H_2	N_2	O_2	CH_4	CO	CO_2	H_2O
Permeance, GPU [a]	300	1	10	0.8	2	20	1000

[a] GPU, gas permeation unit, 10^{-6} cm^3 cm^{-2} s^{-1} cm Hg.

All the model assumptions and reactor configurations employed in this work are shown in the Appendix A (Table A3).

3. Process Design

3.1. Data Communication between Proposed Model and Aspen HYSYS

The Aspen HYSYS has various inbuilt equipment models for chemical industry, which could handle casual simulations. However, the commercial simulator lacks models for newly developed unit operations. In Aspen HYSYS, the default membrane gas separation unit is supplied as an example for user-defined extension, in which a pressure drop in the permeate side of the membrane module could not be calculated, causing up to 40% of deviations on predicted results. Additionally, the inbuilt PFR module has poor performance on complex reaction networks and convergence, which usually takes several minutes to reach convergence; it also has high dependency on initial values, which causes serious convergence problems during the simulation process.

To improve the calculation efficiency and accuracy of the proposed SMR-HT process, this study developed several self-defined extensions for Aspen HYSYS. By communicating with Aspen HYSYS through the COM object in Windows, the data calculated in the Python script in this model could be transferred to Aspen HYSYS [28]. The membrane process was coded and compiled in Visual Basic (VB); by registering the DLL (Dynamic Link Library) file in Aspen HYSYS [29], the simulator could directly calculate the membrane unit with the proposed algorithm. The calculation procedure was coded into a Python script; by performing a programed automation routine, the process in Aspen HYSYS was calculated until convergence. The schematic of the simulation procedure was shown in Figure 3.

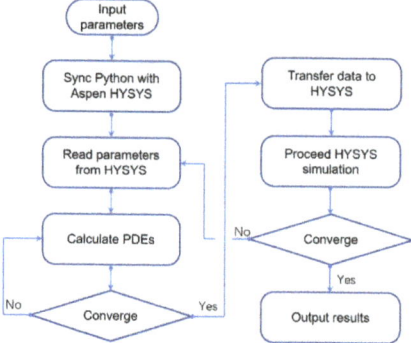

Figure 3. Schematics of simulation procedure of the proposed model.

3.2. Conceptual Design of Hydrogen Production and Pyrolysis Upgrading Process

Hydrogen is usually produced by SMR in a chemical plant or refineries. A typical SMR process includes natural gas steam reformer, water–gas-shifter and gas purification system (PSA or membrane); the latter system is to remove the CH_4, CO_2 and CO content from the raw syngas. In the pyrolytic oil upgrading process, the oxygen and water content will form CO and CO_2, which are also the major byproduct in the syngas. Therefore, there is no need for the gas purification system to remove CO, which enables the system to employ a less expensive unit for gas purification.

Several studies have investigated the feasibility of employing syngas as a hydrogen supplier for the upgrading process. It has been proved in the experiments that the upgrading performance of the pyrolytic oil was almost the same when the partial pressure of hydrogen was similar; some studies proposed that the in-situ reaction of the WGS could promote the upgrading reactions. The results of [30] showed that under the same initial conditions (hydrogen partial pressure), the conversion of the upgrading process was higher in the batch reactor than in continuous flow reactor, proving the impacts of in-situ WGS reactions.

On the basis of that research, this study proposed a novel pyrolytic oil upgrading process with integrated system of a steam methane reformer, gas separation system and HT reaction system. The proposed process utilized a steam methane reformer to produce syngas without a WGS reactor; the hydrogen concentration in the syngas was adjusted by gas separation system, in which the flash gases of the hydrotreating reactor were also purified to produce pure hydrogen as byproduct; the syngas was input into the packed-bed hydrotreating reactor along with pyrolytic oil, and the flashed gases were treated by several steps for hydrogen recovery. The process design is shown in the flow diagram in Figure 4. Noting that the feed flow rate of natural gas was set at a constant of 34.1×10^3 Nm^3/h, under this configuration, the conversion and hydrogen production could be compared and evaluated properly.

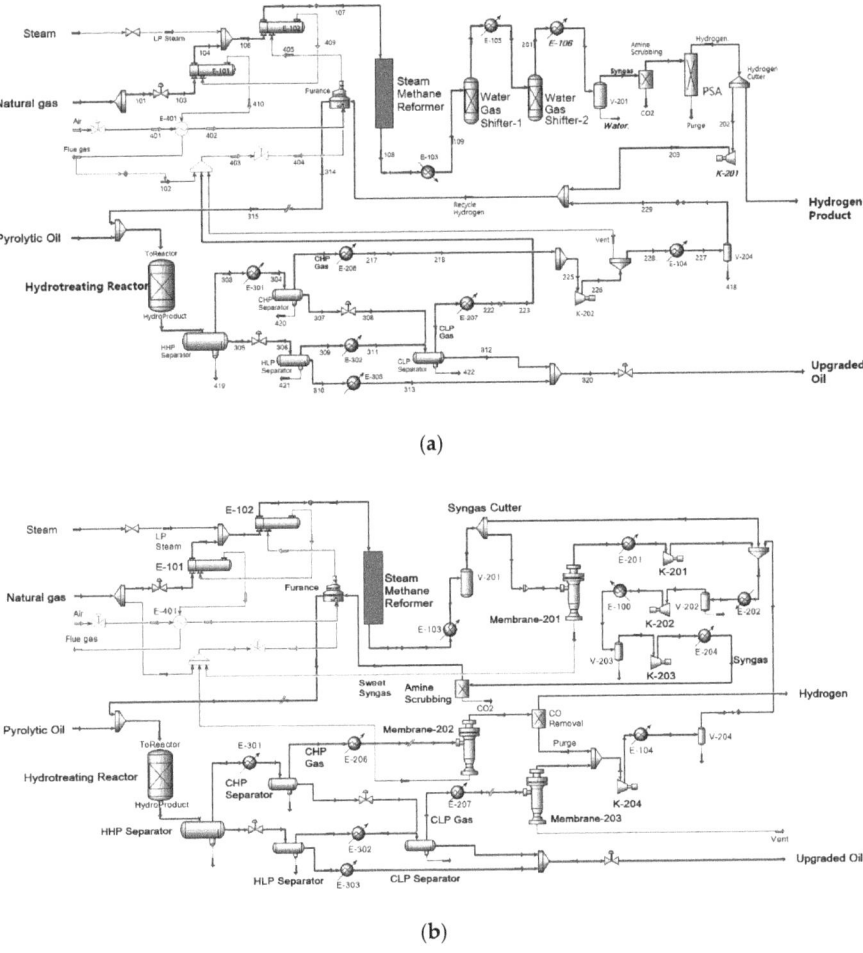

Figure 4. Process flow diagram of pyrolytic oil upgrading process. (**a**) Conventional process with a WGS reactor and PSA unit; (**b**) integrated steam methane reforming-hydrotreating process without a WGS reactor.

The gas separation system in the proposed process was composed by two major parts. The syngas cutter cut part of the stream to a membrane unit (membrane-201), which could adjust the hydrogen concentration by changing membrane area; the residue of membrane-201 was injected into the furnace to recover heat. Another key part for the gas separation system was the flash gas separator (membrane-202), which recovered the hydrogen in the cold high-pressure gas (CHP Gas); with an optional CO removal unit, the permeate from membrane-202 could be transferred to other unit as pure hydrogen product.

The proposed process utilized the CO component in the syngas to induce an in-situ WGS reaction in the hydrotreating reactor, and integrated the pyrolytic oil upgrading process with SMR and gas separation system, which improved the integrity of the plant by merging unit operations, and raised the synthetic utilization of hydrogen and other utilities. The mutual contribution of those aspects would reduce operational expense (OPEX) and capital expense (CAPEX).

4. Discussion

4.1. Reactor Sizing

The core equipment of the pyrolytic oil upgrading process was the hydrotreating reactor. The space velocity of the packed-bed reactor directly determined the performance, which is also the key parameter for scale-up designs of industrial reactors. The impacts of the reactor size on space velocity are shown in the Supporting Information (Figure S1). When the reactor diameter was 3–5 m, and the bed height was higher than 30 m, the space velocity was 0.1–2 h^{-1}, which could satisfy the requirements of a normal HT process.

The performance of the pyrolytic oil upgrading process could be monitored by the conversion of the heavy distillate components. Figure 5 shows the impacts of reactor size on conversion. It could be observed that when the bed length was higher than 80 m and diameter greater than 5 m, the conversion of pyrolytic oil could reach 80% and more. However, higher conversion implied higher risk of a coking problem, and two reactors in series would be required to reach the space velocity, which would increase the CAPEX; the coking problem would induce coke formation and blockage in catalyst pores, causing serious engineering problems. Therefore, choosing an intermediate reactor size would be more suitable for the upgrading process.

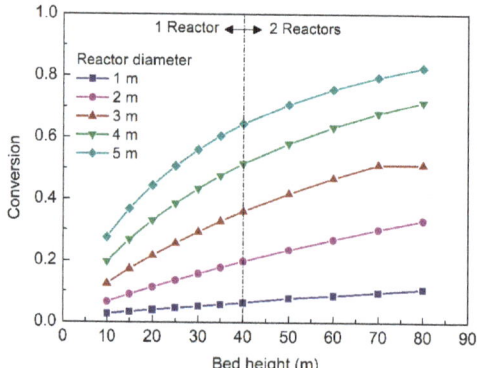

Figure 5. Impacts of bed height on pyrolytic oil conversion.

Another important aspect of the HT process was the hydrogen consumption. Normally, higher conversion of pyrolytic oil leads to higher hydrogen consumption. Figure 6 shows that the space velocity affected the hydrogen consumption greatly (noting that the hydrogen consumption was calculated by weight ratio with respect to the mass flow rate of pyrolytic oil); with increasing reactor size, the hydrogen consumption increased. The cost of hydrogen usually takes 90%–95% of the operational cost.

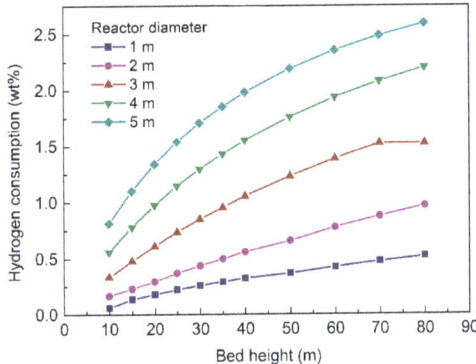

Figure 6. Impacts of bed height and reactor diameter on hydrogen consumption.

Considering the counterbalance demonstrated in Figures 3–5, a reactor with 40 m height and 3 m diameter was chosen as the basis for further simulations to balance the trade-off between reactor size and upgrading performance.

4.2. Effects of Operating Temperature on Pyrolysis Upgrading

The operating temperature of the reactor is crucial for the upgrading process. Figure 7 shows the impacts of operating temperature on the product distribution and the WGS reaction equilibrium constant. It could be observed from the figure that the conversion increased with increasing temperature, along with higher yield of the aromatics lump. Because the WGS reaction is exothermic, higher operating temperatures depressed the reaction equilibrium, and hence reduced the reaction equilibrium constant.

Besides the impacts on yields, a higher operating temperature would produce more coke; although in this study the reaction kinetics did not present much coke or gas formation, the uncertainty of it would be a major consideration for a realistic process. Considering all the impacts of operating temperature, a moderate temperature of 380 °C was selected as the base case for further investigation.

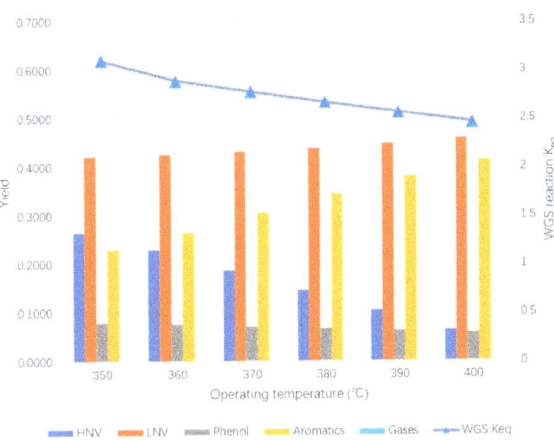

Figure 7. Impacts of operating temperature on product distribution. WGS refers to water–gas-shift reaction; K_{eq} refers to the reaction equilibrium constant.

4.3. Impacts of the Gas Separation System

The gas separation and purification system were the central functional subsystem for the SMR-HT integration process. The gas separation system was composed of three membrane modules. Membrane-201 was employed to enrich the hydrogen in raw syngas; membrane-202 was employed to purify the cold high-pressure flash gas; and membrane-203 was to recover hydrogen from cold low-pressure flash gas. Despite of the impacts of membrane areas, the ratio of syngas cutter was also a determinate factor for the upgrading performance. The mole fraction of hydrogen in the syngas determined the hydrogen partial pressure in the reaction system, and hence determined the reaction process. Therefore, the separation performance of the gases would also provide influences on conversion and hydrogen consumption.

The effects of membrane areas on the syngas hydrogen mole fraction is shown in Figure 8a,b; in which it demonstrated that a lower membrane area in membrane-202 would provide higher hydrogen concentration in the syngas. Figure 8a shows that, the syngas concentration became more sensitive at a higher ratio of syngas cutter. Although the membrane-202 unit did not process syngas directly, its residue was purged into the gas separation system, affecting the syngas concentration indirectly. Comparing Figure 8a,b, when the membrane area of membrane-202 was raised, the hydrogen concentration decreased, implying higher hydrogen recovery.

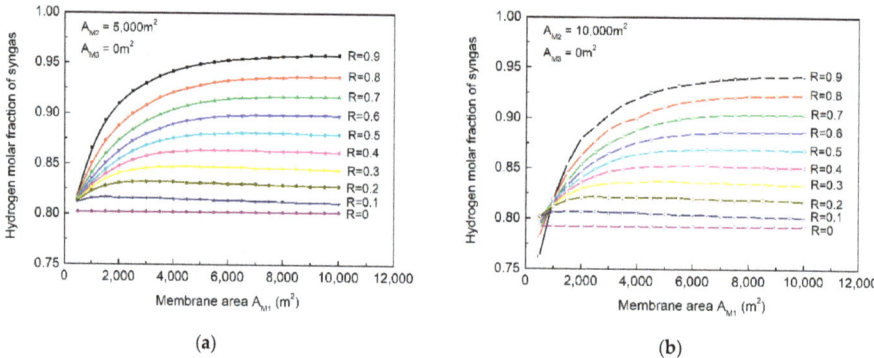

Figure 8. Impacts of syngas cut ratio and membrane area on syngas hydrogen purity. (**a**) Area of membrane-202 was 5000 m^2; (**b**) area of membrane-202 was 10,000 m^2.

The influences of membrane area and syngas cut ratio on conversion are shown in Figure 9a,b. Lower membrane area for membrane-201 and higher cut ratio would significantly reduce the flow rate of raw syngas, which would cease the hydrogenation reactions due to a shortage of reactant (hydrogen). When the membrane area (membrane-201) reached 4000 m^2, and the cut ratio was lower than 0.5, the conversion could reach 0.60. Comparing Figure 9a,b, the increment of membrane area in membrane-202 had little influence on the pyrolytic oil conversion. This was caused by the excess hydrogen in the reaction system. When membrane area and cut ratio were adequate, the reaction progress would be dependent only on hydrogen concentration (partial pressure) at the same temperature.

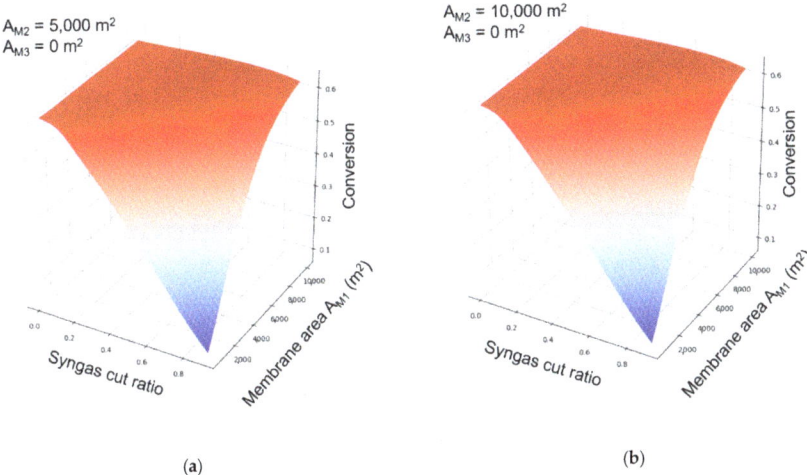

Figure 9. Impacts of membrane area and syngas cut ratio on conversion. (**a**) Area of membrane-202 was 5000 m^2; (**b**) area of membrane-202 was 10,000 m^2.

The impacts of membrane areas on hydrogen consumption are demonstrated in Figure 10a,b, in which it could be observed that with the increasing membrane area in membrane-201, the hydrogen consumption increased. Comparing Figures 10a and 6, it could be deduced that under the same conversion (60% for example), the process configurations with a gas separation system consumed much less hydrogen than simply reducing space velocity (raising reactor diameter and length). This was because of the higher concentration of hydrogen in the syngas raised the partial pressure and boosted the HT reactions, and therefore enable the reaction system to achieve higher conversion with lower hydrogen consumption. Comparing Figure 10a,b, the membrane area of membrane-202 still had little impact on the hydrogen consumption, which was similar to previous discussions (Figures 8 and 9).

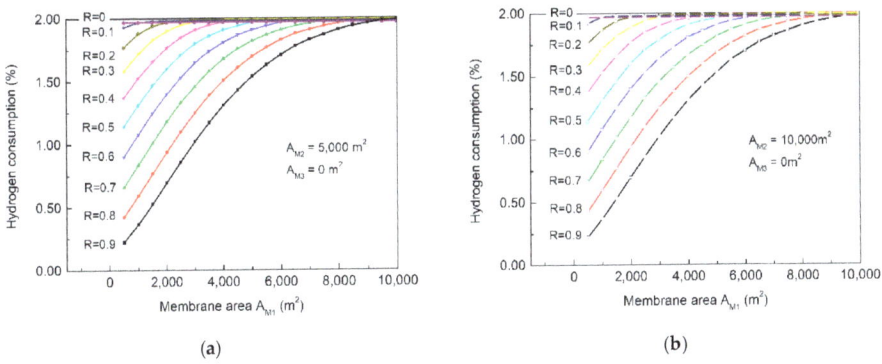

Figure 10. Influences of membrane area on hydrogen consumption. (**a**) Area of membrane-202 was 5000 m^2; (**b**) area of membrane-202 was 10,000 m^2.

Although the impacts of the membrane area in membrane-202 were at a minimum for the reactions, it determined the flow rate of hydrogen product directly. The gas byproducts of the process were mainly hydrocarbons, CO and CO_2. The membrane unit could reject most CO and hydrocarbons in the residue (as fuel gas for the furnace); with a traditional amine scrubbing or adsorption method, the acid gases, such as CO_2, could be removed, and then produce hydrogen with >99.0 mol% purity.

The influences of membrane areas on the flow rate of hydrogen product are shown in Figure 11. When cut ratio was low, increasing the membrane area of membrane-202 had significant positive impact on hydrogen flow rate; when the ratio reached 0.4–0.5, doubling the membrane area of membrane-202 could only provide a little increase in hydrogen flow rate. It is worth noting that when the membrane area of membrane-201 was about 5000 m^2, the impacts of cut ratio became minimal. From Figure 11 we could find the process configuration which could maximize hydrogen production (0.2–0.3 cut ratio, and 5000/10,000 m^2 membrane area for membrane-201/202). Membrane-203 has little impact on separation and reaction because of its small capacity, and the influences are demonstrated in the Supporting Information (Figure S2).

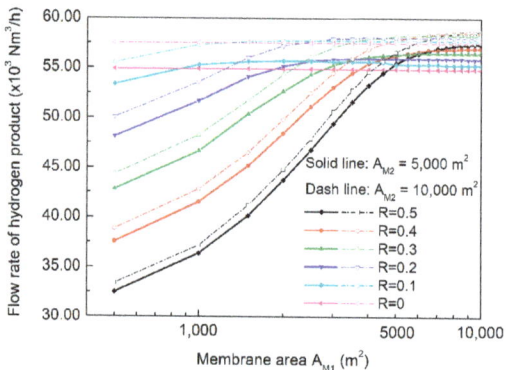

Figure 11. Effects of membrane area and syngas cut ratio on product flow rate of hydrogen.

4.4. Effects of Water Content in Pyrolytic oil

The pyrolytic oil usually contains water content; literatures have reported that the water content could reach 0 wt%–20 wt% in typical pyrolytic oil [31–33]. The water could be utilized as a reactant for the WGS reaction, producing in-situ generated hydrogen which could promote the HT reaction.

This study investigated the effects of water content on pyrolytic oil upgrading. The range that was investigated in this study was 0 wt%–15 wt%, and the kinetics for the water–gas-shift reaction was discussed in Section 2.1. The contributions of the WGS reaction converts CO and water into hydrogen and CO$_2$. The effects are shown in Figure 12; it could be observed that with the increasing water content in the pyrolytic oil, the H$_2$O/CO ratio (molar) at the reactor inlet gradual increased; when water content reached 15 wt%, the H$_2$O/CO ratio was 2.0; according to the stoichiometry of the WGS reaction, the water for the shift reaction would be in excess.

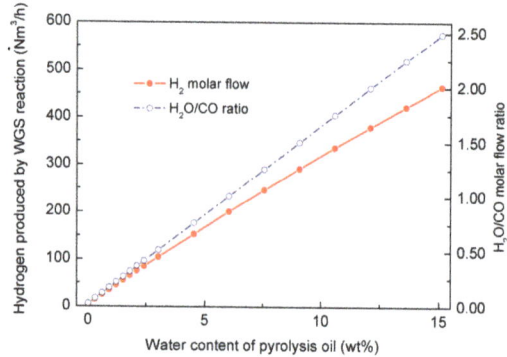

Figure 12. Effects of water–gas-shift reaction on hydrogen production.

Although excess water could promote the WGS reaction, it could also dilute the reactant oil, and hence reduce the conversion. Figure 13 investigated the impacts of water content on conversion and hydrogen consumption; from which it could be observed that with the increasing water content, the conversion gradually decreased, while the hydrogen consumption was decreased by 0.05 percent (with respect to flow rate of pyrolytic oil), which was about 5950 kg/h. The reduced hydrogen consumption would save about 12,000 $/h, and the lower conversion might only cause 200–500 $/h reduction in profit. The WGS reaction could also consume CO, which would benefit the purification process downstream. On the basis of the discussion, it could be concluded that moderate content of water that existed in the pyrolytic oil could reduce hydrogen consumption by promoting the WGS reaction, and hence reduce CO content and OPEX for hydrogen production.

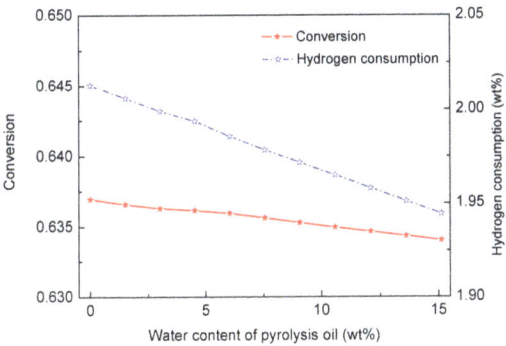

Figure 13. Effects of water content on conversion and hydrogen consumption.

4.5. Techno-Economic Assessment

The evaluations of economics were carried out based on the results of previous sections. When the cut ratio of syngas cutter was 0.4, the upgrading process could achieve higher conversion and hydrogen production. Figure 14 showed the impacts of the cut ratio on the CAPEX of the membrane and compressor (in which the cost was estimated by hourly depreciation rate; only membrane and compressor CAPEX were compared since the two are the major variables for the investment in Figure 14). The CAPEX of the membrane, compressor, reactor, and other key equipment are shown in the Appendix A.

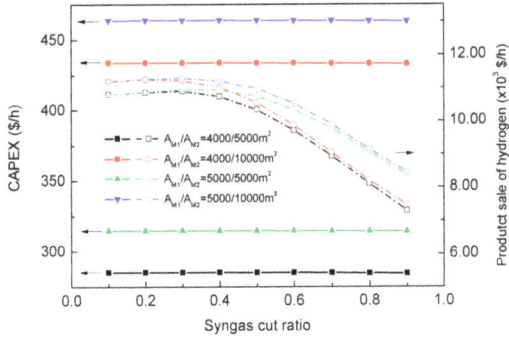

Figure 14. Evaluations on capital expense (CAPEX) and hydrogen product sale.

In Figure 14, the configuration of 0.4 cut ratio and 5000/10,000 m² membrane area (for membrane-201/202 respectively) could provide the highest hydrogen sale; however, the CAPEX for that configuration was significantly higher than others. When the membrane area was reduced to 5000/5000 m² with 0.4 cut ratio, the CAPEX was reduced by 32.0%, while the product sale of hydrogen was only reduced by 2.7%. Considering the almost-identical conversion of the two configurations, the optimal configuration of the proposed SMR-HT process should be 5000/5000/500 m² for membrane-201/202/203 with 0.4 ratio for syngas cutter.

The evaluations of the inlet/outlet throughput and utility consumptions are listed in Table 5 (mass balance of the SMR-HT process is presented in the Supporting Information, Table S1). It could be found that the process consumed significant amount of cooling water, which implies that the heat exchanging network still required further optimization. Nonetheless, Table 5 shows that the proposed integrated SMR-HT process could upgrade pyrolytic oil with high conversion and operational flexibility, while producing hydrogen as a byproduct with low expense; the comprehensive utilization of hydrogen in the proposed process achieved a high hydrogen recovery of 95.6 %, which demonstrated the efficiency of the designed system (reactor size: 5 m diameter and 40m length; membrane area: 5000/5000/500 m² for membrane-201/202/203; ratio for syngas cutter: 0.4).

Table 5. Process evaluations.

Name	Description	Unit	Conventional	SMR-HT Process
Overview	Conversion	%	51.6	63.7
	Hydrogen Consumption	wt%	2.0	2.0
	Hydrogen Recovery	%	90.3	95.6
Inlet	Natural Gas	10^3 Nm³/h	34.1	34.1
	Fuel Natural Gas	10^3 Nm³/h	5.6	0
	Pyrolytic Oil	t/h	119.0	
	Steam	t/h	119.8	
Outlet	Upgraded Oil [a]	t/h	87.5	90.3
	Hydrogen	10^3 Nm³/h	51.5	50.4
	Flue Gas	10^3 Nm³/h	79.1	149.6
	Vent	10^3 Nm³/h	5.9	0.3
	Waste Water	t/h	80.7	88.5
	CO_2	10^3 Nm³/h	25.1	8.7
Utilities	Electricity	kW	2,987	29,672
	Water	t/h	2430.5	1832.3
	Air	10^3 Nm³/h	70.5	128.9

[a] Product upgraded oil was calculated by summing the mass flow of light non-volatile (LNV) + aromatics + phenol.
A conventional SMR plant contains main equipment including a furnace, SMR reactor, WGS reactor, and a separation system (PSA). In the proposed SMR-HT process, the WGS reactor and pressure swing adsorption were substituted by a membrane gas separation system, which could reduce the equipment CAPEX by 30.2%; although the membrane system required intensive compression power, which increased the equipment cost of SMR-HT process by 46.2%, as Table 6 presents.

Table 6. Economic evaluations.

Name	Description	Unit	Conventional	SMR-HT Process
Equipment	SMR Reactor		3.7	3.7
	WGS Reactor		5.6	-
	Hydrotreater	10^6 $	2.4	2.4
	Membrane		-	10.5
	PSA		12.1	-
	Compressor		2.2	21.4

Table 6. Cont.

Name	Description	Unit	Conventional	SMR-HT Process
	Sum of Equipment	10^6 $	26.0	38.0
Feedstock	Natural Gas		12.3	12.3
	Fuel Natural Gas	10^3 $/h	2.0	0
	Pyrolytic Oil			41.7
	Steam			7.2
	Sum of Feedstock	10^3 $/h	63.2	61.2
Product	Upgraded Oil[a]	10^3 $/h	56.9	58.7
	Hydrogen		11.5	11.3
	Sum of Product	10^3 $/h	68.4	70.0
Net profit[b]		10^3 $/h	5.0	8.5

[a] Product upgraded oil was calculated by summing the mass flow of LNV + aromatics + phenol. [b] The net profit was calculated by balancing the feed cost, product sale and equipment depreciation; installation, contingency, labor and maintenance cost were not considered for simplicity. The costs for the equipment and prices for materials were shown in Tables A1 and A2.

However, higher equipment investment of the SMR-HT process contributed to higher conversion and less consumption. In the comparisons of feedstock and product sale sections of Table 6, it could be concluded that the SMR-HT process not only reduced additional fuel gas (natural gas) for heat balance, the high conversion also contributed greatly on product sale. When equipment depreciation cost was considered, the net profit showed that the SMR-HT process could provide 70.0% higher hourly profit than the conventional process. Note that the membrane unit was calculated by an excessively high price (1000 $/m^2); while in other studies, the membrane unit usually possessed low price as 50–500 $/m^2 [34,35]. The purpose of the high price in this study was to counterbalance the possibility of overestimation: Housing, pipes, unscheduled shutdown of compressors. The high price could compensate for those factors to some extent and provide reasonable results for comparisons. In the research of Ohs et al. [36], it indicated that the housing of membrane material took the major part of the investment, which implies that after four years of usage, the replacement of membrane material would cost much less, giving advantages for the proposed SMR-HT process.

5. Conclusions

In this study, the integrated process of steam methane reforming and pyrolytic oil upgrading process enhanced by a gas separation system was proposed conceptually. The integrated process substituted the WGS reactor with a membrane gas separation unit, allowing the system to adjust hydrogen concentration in the syngas; the pyrolytic oil was hydrotreated by syngas, which reduced the OPEX on hydrogen consumption. A synthesis gas recovery system was designed to recover hydrogen from the flash gases, which could improve the utilization of hydrogen and further produce pure hydrogen as product.

A simulation framework was developed to model the hydrotreater and membrane module. The hydrotreater was modeled by an axial-dispersion model; five-lump reaction kinetics were employed to simulate the upgrading process of the pyrolytic oil. The proposed framework provided data communications between self-defined algorithm and commercial simulator Aspen HYSYS, enabling the simulator to solve complex reaction kinetics and optimize the process in a more efficient way.

The optimizations of the proposed process aided it to achieve 63.7% conversion with 2.0 wt% hydrogen consumption and 95.6% hydrogen recovery. The effects of water content was investigated, and the results showed that moderate water content could promote the in-situ WGS reaction, and improve the hydrogen production with slight reduction in conversion. The proposed integrated process enabled the upgrading process to achieve higher performance with simplified design and flexible operating. The results demonstrated that the proposed integrated process could upgrade pyrolytic oil and produce hydrogen synthetically, providing 70% higher net profit than the conventional process.

Supplementary Materials: The following are available online at http://www.mdpi.com/2227-9717/7/5/284/s1. Figure S1: Liquid special hourly velocity of pyrolytic oil hydroprocessing reactor. Figure S2: Impacts of membrane area (Membrane-203) on conversion and hydrogen consumption. Table S1: Mass balance of SMR-HT process. The reactor sizing and impacts of the membrane area of Membrane-203 were provided in the supplementary file.

Author Contributions: Conceptualization, B.C. and T.Y.; investigation, B.C. and W.X.; software, B.C. and W.X.; visualization, B.C.; writing—original draft, B.C.; writing—review and editing, T.Y.; W.X. and A.k.N.

Funding: National Natural Science Foundation of China (Grant Nos. 61627803 and 61671319) and National Key R&D Program of China (Grant Nos. 2018YFA0209400 and 2018YFA0209404).

Acknowledgments: The authors would like to thank the financial supports from the National Natural Science Foundation of China (Grant Nos. 61627803 and 61671319) and the National Key R and D Program of China (Grant Nos. 2018YFA0209400 and 2018YFA0209404).

Conflicts of Interest: The authors declare no conflict of interest.

Appendix A

The CAPEX of key equipment, such as reactors, membrane unit and compressor are shown in this section. The depreciation rates of the membrane and PSA were four and eight years respectively (membrane material needs to be replaced every 4–6 years due to aging problems; the adsorbents of PSA unit should be substituted every 4–8 years because of degrading); other equipment, i.e., compressor and reactors, are calculated on a basis of 20 years depreciation.

The parameters provided in Table A1 are the average investment cost with respect to unit throughput (expected membrane unit was calculated by area).

Table A1. Investment cost for key equipment.

Equipment	Unit	CAPEX
Membrane	$/m^2	1000
Compressor	$/kW	720
SMR Reactor	10^4 $/($10^3$ Nm^3 h^{-1} Unit)	2.0
WGS Reactor	10^4 $/($10^3$ Nm^3 h^{-1} Unit)	1.2
Hydrotreating Reactor	10^6 $/($10^6$ t a^{-1} Unit)	2.4
PSA	10^5 $/($10^3$ Nm^3 h^{-1} Unit)	1.1

Table A2. Prices for oil and gases.

Name	Unit	Cost or Price
Natural Gas	$/t	500
Pyrolytic Oil	$/t	350
Upgraded Oil	$/t	650
Hydrogen	10^3 $/t	2.5
Steam	$/t	60

The assumptions in the simulation works of this study are listed in the following table (Table A3).

Table A3. Assumptions for modeling.

Name	Description	Mechanism/Configuration
Membrane Gas Separation	• Module form: • Mass transfer: • Membrane permeation: • Flow pattern: • Pressure drop:	Hollow-fiber Plug flow Solution-diffusion Counter-current Hagen–Poiseuille

Table A3. *Cont.*

Name	Description	Mechanism/Configuration
Steam Methane Reformer	• Reactor form: • Mass transfer: • Rate constant:	Tube reactor Plug flow Arrhenius and equilibrium
Water Gas Shifter	• Reactor form: • Mass transfer: • Rate constant:	Bubble bed reactor Plug flow Arrhenius and equilibrium
Hydrotreater	• Reactor form: • Mass transfer: • Rate constant:	Packed bed reactor Axial dispersion Arrhenius

References

1. Guedes, R.E.; Luna, A.S.; Torres, A.R. Operating parameters for bio-oil production in biomass pyrolysis: A review. *J. Anal. Appl. Pyrolysis* **2018**, *129*, 134–149. [CrossRef]
2. Dhyani, V.; Bhaskar, T. A comprehensive review on the pyrolysis of lignocellulosic biomass. *Renew. Energy* **2018**, *129*, 695–716. [CrossRef]
3. Gómez, N.; Banks, S.W.; Nowakowski, D.J.; Rosas, J.G.; Cara, J.; Sánchez, M.E.; Bridgwater, A.V. Effect of temperature on product performance of a high ash biomass during fast pyrolysis and its bio-oil storage evaluation. *Fuel Process. Technol.* **2018**, *172*, 97–105. [CrossRef]
4. Kan, T.; Strezov, V.; Evans, T.J. Lignocellulosic biomass pyrolysis: A review of product properties and effects of pyrolysis parameters. *Renew. Sustain. Energy Rev.* **2016**, *57*, 1126–1140. [CrossRef]
5. McCormick, R.L.; Ratcliff, M.A.; Christensen, E.; Fouts, L.; Luecke, J.; Chupka, G.M.; Yanowitz, J.; Tian, M.; Boot, M. Properties of Oxygenates Found in Upgraded Biomass Pyrolysis Oil as Components of Spark and Compression Ignition Engine Fuels. *Energy Fuels* **2015**, *29*, 2453–2461. [CrossRef]
6. Banks, S.W.; Nowakowski, D.J.; Bridgwater, A.V. Impact of Potassium and Phosphorus in Biomass on the Properties of Fast Pyrolysis Bio-oil. *Energy Fuels* **2016**, *30*, 8009–8018. [CrossRef]
7. Mullen, C.A.; Boateng, A.A. Mild hydrotreating of bio-oils with varying oxygen content produced via catalytic fast pyrolysis. *Fuel* **2019**, *245*, 360–367. [CrossRef]
8. Ardiyanti, A.R.; Bykova, M.V.; Khromova, S.A.; Yin, W.; Venderbosch, R.H.; Yakovlev, V.A.; Heeres, H.J. Ni-Based Catalysts for the Hydrotreatment of Fast Pyrolysis Oil. *Energy Fuels* **2016**, *30*, 1544–1554. [CrossRef]
9. Iisa, K.; French, R.J.; Orton, K.A.; Dutta, A.; Schaidle, J.A. Production of low-oxygen bio-oil via ex situ catalytic fast pyrolysis and hydrotreating. *Fuel* **2017**, *207*, 413–422. [CrossRef]
10. Gholizadeh, M.; Gunawan, R.; Hu, X.; de Miguel Mercader, F.; Westerhof, R.; Chaitwat, W.; Hasan, M.M.; Mourant, D.; Li, C.Z. Effects of temperature on the hydrotreatment behaviour of pyrolysis bio-oil and coke formation in a continuous hydrotreatment reactor. *Fuel Process. Technol.* **2016**, *148*, 175–183. [CrossRef]
11. Voldsund, M.; Jordal, K.; Anantharaman, R. Hydrogen production with CO_2 capture. *Int. J. Hydrog. Energy* **2016**, *41*, 4969–4992. [CrossRef]
12. Golmakani, A.; Fatemi, S.; Tamnanloo, J. Investigating PSA, VSA, and TSA methods in SMR unit of refineries for hydrogen production with fuel cell specification. *Sep. Purif. Technol.* **2017**, *176*, 73–91. [CrossRef]
13. Marques, J.P.; Matos, H.A.; Oliveira, N.M.C.; Nunes, C.P. State-of-the-art review of targeting and design methodologies for hydrogen network synthesis. *Int. J. Hydrog. Energy* **2017**, *42*, 376–404. [CrossRef]
14. Fu, Y.C.; Illig, E.G. Catalytic Coal Liquefaction Using Synthesis Gas. *Ind. Eng. Chem. Process Des. Dev.* **1976**, *15*, 392–396. [CrossRef]
15. Kumar, A.; Kumar, N.; Baredar, P.; Shukla, A. A review on biomass energy resources, potential, conversion and policy in India. *Renew. Sustain. Energy Rev.* **2015**, *45*, 530–539. [CrossRef]
16. Saidur, R.; Abdelaziz, E.A.; Demirbas, A.; Hossain, M.S.; Mekhilef, S. A review on biomass as a fuel for boilers. *Renew. Sustain. Energy Rev.* **2011**, *15*, 2262–2289. [CrossRef]
17. Zacher, A.H.; Olarte, M.V.; Santosa, D.M.; Elliott, D.C.; Jones, S.B. A review and perspective of recent bio-oil hydrotreating research. *Green Chem.* **2014**, *16*, 491–515. [CrossRef]

18. Gollakota, A.R.K.; Subramanyam, M.D.; Kishore, N.; Gu, S. CFD simulations on the effect of catalysts on the hydrodeoxygenation of bio-oil. *RSC Adv.* **2015**, *5*, 41855–41866. [CrossRef]
19. Muhlbauer, A. *Phase Equilibria: Measurement & Computation*; CRC Press: Boca Raton, FL, USA, 1997.
20. Stowe, L.R. Method of Conversion of Heavy Hydrocarbon Feedstocks. U.S. Patent 5,547,563, 20 August 1996.
21. Wang, G.; Li, W.; Li, B.; Chen, H. Direct liquefaction of sawdust under syngas. *Fuel* **2007**, *86*, 1587–1593. [CrossRef]
22. Guyer, J.E.; Wheeler, D.; Warren, J.A. FiPy: Partial differential equations with Python. *Comput. Sci. Eng.* **2009**, *11*, 6–15. [CrossRef]
23. Chen, B.; Dai, Y.; Ruan, X.; Xi, Y.; He, G. Integration of molecular dynamic simulation and free volume theory for modeling membrane VOC/gas separation. *Front. Chem. Sci. Eng.* **2018**, *12*, 296–305. [CrossRef]
24. Chen, B.; Ruan, X.; Jiang, X.; Xiao, W.; He, G. Dual-Membrane Module and Its Optimal Flow Pattern for H_2/CO_2 Separation. *Ind. Eng. Chem. Res.* **2016**, *55*, 1064–1075. [CrossRef]
25. Chen, B.; Ruan, X.; Xiao, W.; Jiang, X.; He, G. Synergy of CO_2 removal and light hydrocarbon recovery from oil-field associated gas by dual-membrane process. *J. Nat. Gas Sci. Eng.* **2015**, *26*, 1254–1263. [CrossRef]
26. Wang, L.; Zhang, Y.; Wang, R.; Li, Q.; Zhang, S.; Li, M.; Liu, J.; Chen, B. Advanced monoethanolamine absorption using sulfolane as a phase splitter for CO_2 capture. *Environ. Sci. Technol.* **2018**, *52*, 14556–14563. [CrossRef]
27. Wang, R.; Liu, S.; Wang, L.; Li, Q.; Zhang, S.; Chen, B.; Jiang, L.; Zhang, Y. Superior energy-saving splitter in monoethanolamine-based biphasic solvents for CO_2 capture from coal-fired flue gas. *Appl. Energy* **2019**, *242*, 302–310. [CrossRef]
28. Chen, B.; Meng, Z.; Ge, H.; Alcheikhhamdon, Y.; Hoorfar, M.; Liu, L.; Yang, T.; Fang, X. Optimization of Residual Oil Hydrocrackers: Integration of Pump-free Ebullated-bed Process with Membrane-aided Gas Recovery System. *Energy Fuels* **2019**, *33*, 2584–2597. [CrossRef]
29. Hoorfar, M.; Alcheikhhamdon, Y.; Chen, B. A novel tool for the modeling, simulation and costing of membrane based gas separation processes using Aspen HYSYS: Optimization of the CO_2/CH_4 separation process. *Comput. Chem. Eng.* **2018**, *117*, 11–24. [CrossRef]
30. Cortes, J.; Valencia, E. Configuration of Adsorbed Phases and Their Evolution to Absorbent States in the CH4–O2 Catalytic Reaction. *Bull. Chem. Soc. Jpn.* **2009**, *82*, 683–688. [CrossRef]
31. Zhang, Q.; Chang, J.; Wang, T.; Xu, Y. Review of biomass pyrolysis oil properties and upgrading research. *Energy Convers. Manag.* **2007**, *48*, 87–92. [CrossRef]
32. Westerhof, R.J.M.; Kuipers, N.J.M.; Kersten, S.R.A.; van Swaaij, W.P.M. Controlling the water content of biomass fast pyrolysis oil. *Ind. Eng. Chem. Res.* **2007**, *46*, 9238–9247. [CrossRef]
33. Czernik, S.; Johnson, D.K.; Black, S. Stability of wood fast pyrolysis oil. *Biomass Bioenergy* **1994**, *7*, 187–192. [CrossRef]
34. Bhide, B.D.; Stern, S.A. Membrane processes for the removal of acid gases from natural gas. II. Effects of operating conditions, economic parameters, and membrane properties. *J. Membr. Sci.* **1993**, *81*, 239–252. [CrossRef]
35. Merkel, T.C.; Zhou, M.; Baker, R.W. Carbon dioxide capture with membranes at an IGCC power plant. *J. Membr. Sci.* **2012**, *389*, 441–450. [CrossRef]
36. Ohs, B.; Lohaus, J.; Wessling, M. Optimization of membrane based nitrogen removal from natural gas. *J. Membr. Sci.* **2016**, *498*, 291–301. [CrossRef]

© 2019 by the authors. Licensee MDPI, Basel, Switzerland. This article is an open access article distributed under the terms and conditions of the Creative Commons Attribution (CC BY) license (http://creativecommons.org/licenses/by/4.0/).

Article

Hydrogen Production and Subsequent Adsorption/Desorption Process within a Modified Unitized Regenerative Fuel Cell

Diksha Kapoor [1], Amandeep Singh Oberoi [2,*] and Parag Nijhawan [1]

1 Electrical and Instrumentation Engineering Department, Thapar Institute of Engineering and Technology Patiala, Punjab-147004, India; dixakapoormail@gmail.com (D.K.); parag.nijhawan@rediffmail.com (P.N.)
2 Mechanical Engineering Department, Thapar Institute of Engineering and Technology Patiala, Punjab 147004, India
* Correspondence: oberoi@thapar.edu

Received: 22 March 2019; Accepted: 22 April 2019; Published: 24 April 2019

Abstract: For sustainable and incremental growth, mankind is adopting renewable sources of energy along with storage systems. Storing surplus renewable energy in the form of hydrogen is a viable solution to meet continuous energy demands. In this paper the concept of electrochemical hydrogen storage in a solid multi-walled carbon nanotube (MWCNT) electrode integrated in a modified unitized regenerative fuel cell (URFC) is investigated. The method of solid electrode fabrication from MWCNT powder and egg white as an organic binder is disclosed. The electrochemical testing of a modified URFC with an integrated MWCNT-based hydrogen storage electrode is performed and reported. Galvanostatic charging and discharging was carried out and results analyzed to ascertain the electrochemical hydrogen storage capacity of the fabricated electrode. The electrochemical hydrogen storage capacity of the porous MWCNT electrode is found to be 2.47 wt%, which is comparable with commercially available AB_5-based hydrogen storage canisters. The obtained results prove the technical feasibility of a modified URFC with an integrated MWCNT-based hydrogen storage electrode, which is the first of its kind. This is surelya step forward towards building a sustainable energy economy.

Keywords: hydrogen energy; solid-state hydrogen storage; unitized regenerative fuel cell; multi-walled carbon nanotube; proton battery

1. Introduction and Background

For many centuries, exhaustible fuels formed the basis of meeting most of the world's energy demands. The continuous increment in the world's population and technological development has led to an extensive and incremental utilization of fossil fuels [1]. The increasing concentrations of the greenhouse gases in the Earth's atmosphere due to human activities has resulted in global warming and related climate change. Burning of fossil fuels is one of the major activities leading to the emission of greenhouse gases [2]. Therefore, these reasons generate a need of transition from fossil fuels to renewable energy sources over the coming decades. Indeed, one of the biggest challenges facing humankind over the next few years is decreasing dependence on fossil fuels (such as natural gas) and their by-products and controlling the emission of greenhouse gases responsible for climate change. In other words, it is necessary to make a shift towards a sustainable energy economy [3].

One of the versatile, sustainable, and scalable forms of energy storage is hydrogen [4,5]. Hydrogen could act as an energy carrier and prove as an alternative to petroleum products [6,7]. Moreover, hydrogen is capable of producing electricity via electrochemical reaction when used in a fuel cell [8]. A fuel cell is an electrochemical device capable of converting the chemical energy of the reactants into electricity [9]. The proton exchange membrane (PEM)-unitized regenerative fuel cell URFC contains a proton exchange membrane as an electrolyte (commonly known as membrane electrode assembly (MEA)) and operates at room temperature. Hydrogen can be stored in the form of gas under high pressure, liquid at cryogenic temperatures, and as a chemical compound known as solid-state or electrochemical storage. Electrochemical hydrogen storage is safer compared to its peer storage forms [10] and has attracted the maximum attention of researchers. However, the best reported figure of electrochemical hydrogen storage is less than U.S. department of energy (DOE) targets. Conventional hydrogen production and storage system involves an electrolyzer, gas compressor, and fuel cell. Having lots of mechanical components results in lower round-the-trip efficiency. Therefore, a URFC was introduced by Bahaman and Andrews in 2015 [11] that involved dependency on a gas storage system and incurred energy expenditure, which became the key limitation of the system. To overcome this limitation of the URFC, a modified URFC was introduced in 2015 by Professor John Andrews and Saeed Seif Mohammadi and they called it as a proton battery [12]. Proton battery is a modified URFC that could run as an electrolyzer to generate hydrogen from water, and store hydrogen in ionic form and fuel cells to give electricity and water. A schematic of charging (electrolysis mode) and discharging (fuel cell mode) of modified URFC or proton battery with hydrogen ion is shown in Figures 1 and 2, respectively.

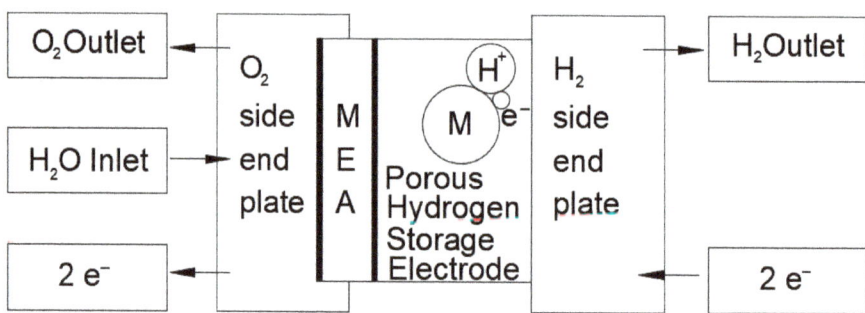

Figure 1. Representing the charging mode or electrolysis mode of the proton battery.

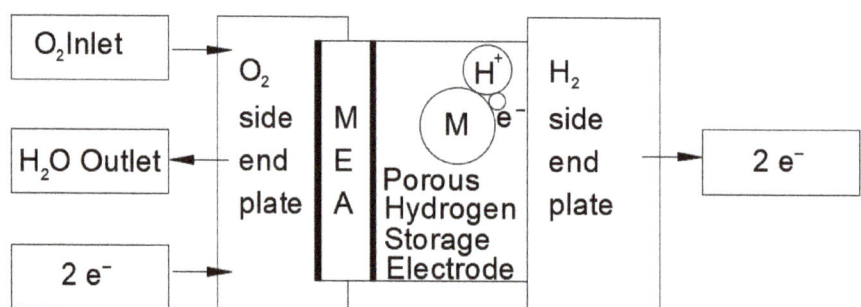

Figure 2. Representing the discharging mode or fuel cell mode of the proton battery.

In the earlier research work, aAB$_5$-type metal hydride electrode was used for the storage of hydrogen within a modified URFC, but the stored hydrogen formed strong chemical bonds with the storage medium and failed to come out [12]. Additionally, AB$_5$ reportedly encourage the formation of hydrogen gas, which is to be discouraged in the operation of a modified URFC [13]. The authors did manage to store 1 wt% of hydrogen in a porous AB$_5$-based electrode, but no significant signs of hydrogen discharge are reported [12]. Moreover, metal hydride being heavy, costly, and having low degree of reversibility were replaced by carbon. In 2013, Javad Jazaeri reported on testing of a composite activated carbon–nafion (aC-nafion) electrode, instead of AB_5 metal hydride, in a modified URFC for electrochemical hydrogen storage [14]. Carbon has a large internal pore surface area, ease of availability, low cost, and is light weight; it has been considered as a suitable alternative for further research work. However, no sign of electrochemical hydrogen storage was reported by Jazaeri with an explanation that nafion (chemically known as perfluorosulfonic acid), which served as the proton conducting medium within the activated carbon electrode, failed to penetrate in aC due to its larger molecular size [14]. Therefore, further research focused on the use of liquid proton conducting mediums within the porous storage electrodes [15]. Heidari S. et al. in 2018 reported on achieving nearly 1 wt% of electrochemical hydrogen storage in an acid-soaked aC electrode during charging and 0.8 wt% during discharging [15]. As per the literature [16–20], multi-walled carbon nanotubes (MWCNT)is a form of carbon that has potential to store hydrogen, but its technical feasibility for electrochemical hydrogen storage within a proton battery or a modified URFC is yet to be proven. The aim of this paper is; therefore, to investigate experimentally the electrochemical hydrogen storage capacity of a solid MWCNT-based electrode when integrated in a modified URFC or proton battery.

2. Experiment and Materials

Details regarding materials used to conduct this experiment have been mentioned in Table 1.

Table 1. List of materials used in this experiment.

S. No.	Product Name	Manufacturer
1	Multi walled carbon nanotube	Platonic Nanotech Private Limited, India
2	Membrane electrode assembly	Saienergy Fuel Cell India Pvt Ltd.
3	Carbon paper and carbon cloth	Saienergy Fuel Cell India Pvt Ltd.
4	Gas collection cylinder	Fabricated at TIET *
5	Bi-polar end plates	Fabricated at TIET
6	DC power supply	scientiFic
7	Electrical load	Fabricated at TIET
8	Die mold	Fabricated at TIET

* Thapar Institute of Engineering and Technology.

2.1. Fabrication of MWCNT Electrode

The specification of MWCNT used in this experiment is listed in Table 2. For fabrication of the electrode, a measured quantity of MWCNT was mixed with 19.25% of egg white (used as binder) to prepare the electrode. The mixture of egg white and MWCNT was mixed well and poured into a mold to attain a desired shape. Egg whites consist of 90% water, which, when evaporated, leaves behind the building material to bind up together the powdered MWCNT particles [21]. It was just enough to bind the MWCNT particles and any excess of the binder could cover-up the pores of MWCNT particles, which are active sited for H-storage. After evaporating the moisture content of the mixture in the sun for 2 days, the electrode was ready, as shown in Figure 3a. The mold used for casting the electrode (shown in Figure 3b) was designed and 3D printed with specifications 25 × 25 × 2 mm.

Table 2. List of Specification related to Multi-walled Carbon Nanotube (MWCNT).

Specification	Unit	Standard
Diameter	Nm	10 to 15 nm
Length	μm	2–10 μm
Ash content	%	<2%
Purity	%	>97%
Specific surface Area	m^2/g	250 to 270 m^2/g
Bulk density	g/cm^3	0.06 to 0 g/cm^3

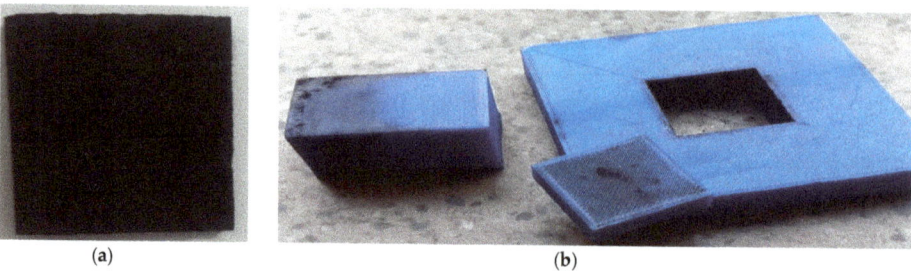

Figure 3. (a) Prepared multi-walled carbon nanotube electrode; (b) 3D mold used for fabrication of electrode.

In case of solid-state hydrogen storage, pores play an important role. The porosity of the electrode can be ascertained from Figure 4, which is a scanning electron microscopy (SEM) image of the MWCNT electrode.

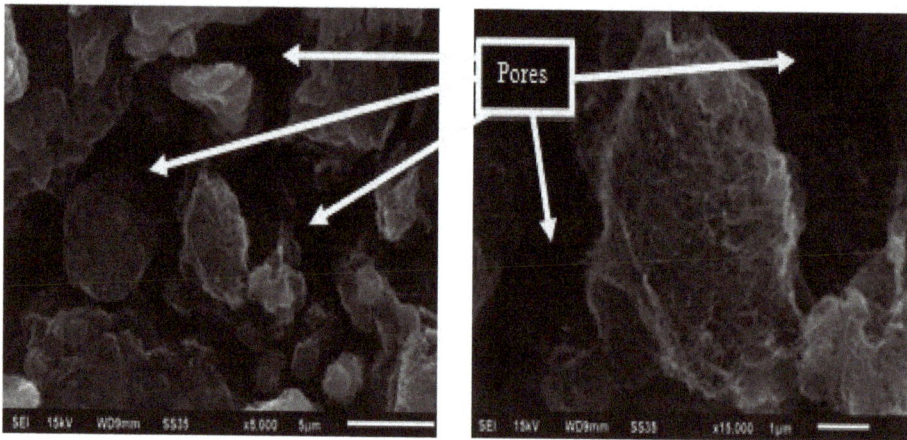

Figure 4. Electron Microscopy (SEM) images of the Multi-walled Carbon Nanotube (MWCNT) electrode.

Pores are classified according to their size [22]. Table 3 classifies different types of pores along with their sizes. The pores are interconnected, as shown in Figure 5, and are the sites for electrochemical hydrogen storage. The major amount of hydrogen stored in solid state is in ultramicropores [23].

Table 3. Types of pore and their size.

Type of Pore	Size
Ultramicropores	<0.7 nm
Micropores	0.7 and 2 nm
Mesopores	2 and 50 nm
Macropores	50 nm

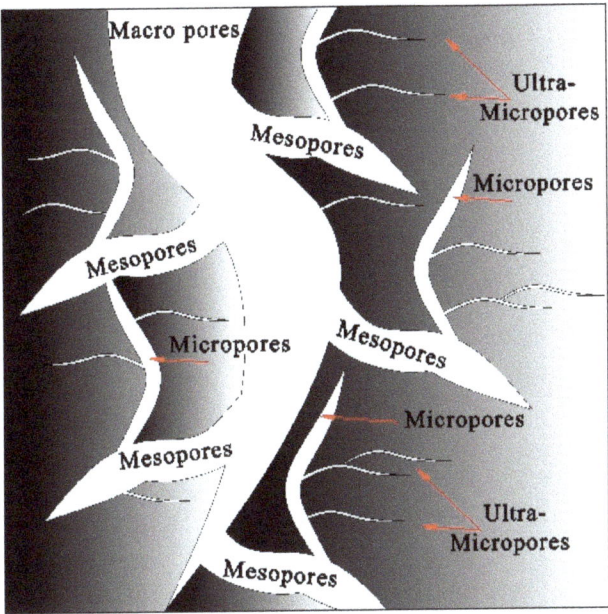

Figure 5. Pictorial description of various pores for electrochemical hydrogen storage.

Egg white, being an organic binder, has a major component of carbon and oxygen. From the EDS (energy dispersive spectroscopy), as mentioned in Table 4, the carbon content of the electrode was found to be 84.16%. As the egg white used was 19.25%, the rest was MWCNT, thus the carbon content of the porous electrode should be more than 80.75%. The presence of other elements was because of the egg white which is an organic protein.

Table 4. Multi-walled Carbon Nanotube (MWCNT) electrode Energy Dispersive Spectroscopy (EDS) result.

Element	Weight (%)	Atomic (%)
C, K	84.16	88.54
O, K	13.28	10.19
Na, K	0.41	0.23
Mg, K	0.30	0.16
Al, K	0.71	0.08
Si, K	0.05	0.02
S, K	0.43	0.17
Cl, K	0.38	0.14
Fe, K	0.82	0.19

2.2. Fabrication of Modified URFC

The modified URFC was fabricated with the constituter parts: membrane electrode assembly, bi-polar end plates, gas diffusion layer, rubber gasket, and MWCNT electrode. Membrane electrode assembly (MEA) consists of a proton exchange membrane and catalyst for the oxygen and hydrogen sides. The presence of a catalyst helps in initiating the breakdown of water at the anode side during charging (electrolysis mode), and the release of electron along with proton from the electrode during discharging (fuel cell mode). In this experiment, the MEA was a reversible type with catalyst loading on the oxygen side as iridium oxide plus platinum black, and on the hydrogen side the loading was of platinum black. Nafion 117 was used as a proton exchange membrane in modified URFC.

On performing SEM test on MEA (shown in Figure 6), it was found that catalyst loading on MEA was evenly distributed, which favors the controlled water dissociation and hydrogen redox reaction. The active area of MEA was 25 mm × 25 mm, which was equal to the size of the electrode. In order to check the presence of impurity in MEA, EDS test was performed on the anode and cathode sides, as mentioned in Tables 5 and 6, respectively. No substantial impurity was found on both sides of MEA.

Figure 6. Scanning Electron Microscopy (SEM) image of membrane electrode assembly (MEA) depicting even catalyst loading.

Table 5. Energy Dispersive Spectroscopy (EDS) result of Membrane Electrode Assembly (MEA) for anode (oxygen) side.

Element	Weight (%)	Atomic (%)
C, K	41.63	66.95
O, K	15.08	15.03
F, K	15.16	14.65
Si, K	0.64	0.44
Fe, K	1.15	0.4
Ir, M	11.72	1.60
Pt, M	15.35	1.68

Table 6. Energy Dispersive Spectroscopy (EDS) result for membrane electrode assembly (MEA) cathode (hydrogen) side.

Element	Weight (%)	Atomic (%)
C, K	56.82	74.72
O, K	10.22	10.09
F, K	15.49	12.88
Si, K	0.47	0.26
S, K	1.65	0.81
Pt, M	15.36	1.24

The bi-polar end plates of modified URFC, shown in Figure 7, are generally made up of metals such as stainless steel, titanium, or metallic alloys. The end plates should have high thermal conductivity, high electrical conductivity, high resistance to corrosion, low permeability to gas, high mechanical strength, and low mass. Stainless steel metallic (SS316) plates are widely used for research activity because they have low cost, ease of manufacturing, high thermal and electrical conductivity, and good mechanical properties. The end plate contains header and flow channels that provide path for oxygen flow to the oxygen side of the membrane and the removal of hydrogen to other side. The flow channels present on both the anode and cathode side end plates were of serpentine design and were identical except for the number openings in the headers. The oxygen side end plate or anode end plate consisted of three openings in the header—one for water inlet, one for oxygen gas outlet, and the third one for oxygen gas inlet, required while discharging the cell. On the other hand, the hydrogen side end plate or cathode end plate had two openings in the header—one for the hydrogen gas outlet and another for the hydrogen gas inlet, if required to enhance the reaction rate while discharging. The active area for activated carbon and electrode assembly was 25 × 25 mm. The end plates on both sides had the same dimensions (i.e., 72 × 84 mm), but varied in thickness. End plate on the hydrogen side was 9 mm thick to accommodate the electrode, whereas the oxygen side end plate was 7 mm thick. The flow channel plays an important role in proper and even circulation of reactant gases on the electrode, as well as in the removal of water from the electrode [24]. Figure 7 shows the oxygen and hydrogen side end plates along with their flow channels.

Figure 7. Actual hydrogen and oxygen side end plates of the modified unitized regenerative fuel cell URFC.

Gas diffusion layer plays important part in a modified URFC. It provides the function of: MEA mechanical support, allowing diffusion of gas in and out of active areas, and protecting the catalyst from erosion and corrosion. Carbon cloth (CC) has been used as gas diffusion layer (GDL) for the hydrogen and oxygen sides of the electrode. The active area of carbon cloth was 2.5 × 2.5 cm^2 same as that of the porous electrode and MEA.

2.3. Testing

To test MWCNT electrode's electrochemical hydrogen storage capacity, modified URFC was assembled as shown in Figure 8. The porous electrode was soaked in 0.2 mL of 1M dilute sulfuric acid (dilute); this was done in order to provide a conductive medium for protons inside the MWCNT electrode. The MEA and rubber gasket was sealed with silicone gel in order to avoid any leakage of gas. For electrical separation of the oxygen and hydrogen sides, nut and bolt assembly was covered with insulating sleeves along with plastic washers.

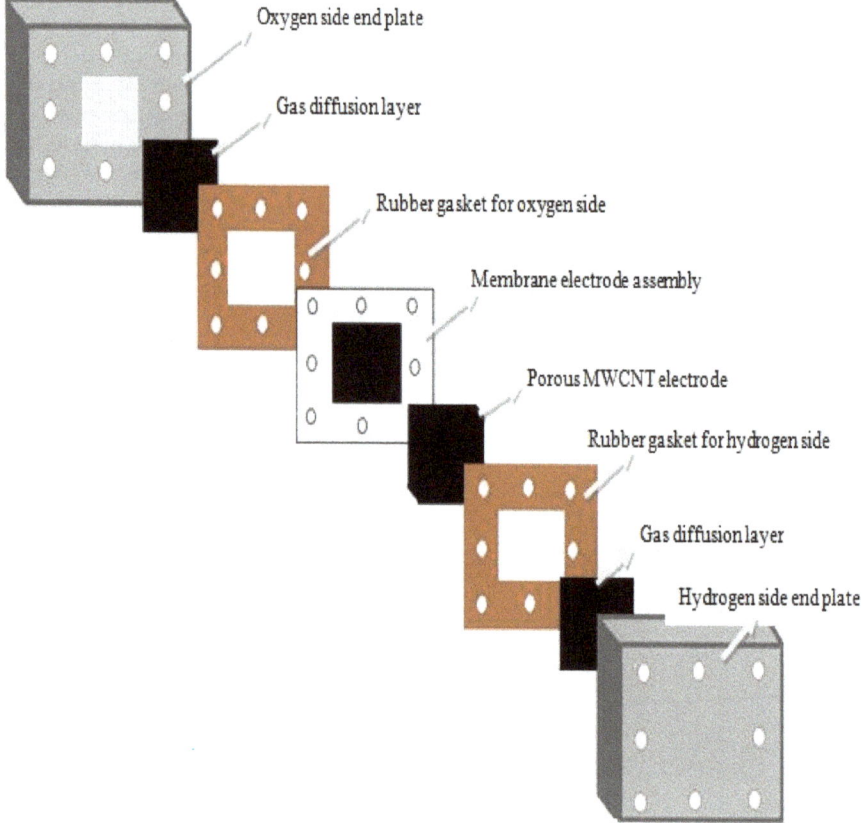

Figure 8. Schematic diagram of orientation for a modified unitized regenerative fuel cell URFC.

The setup also included gas collection cylinders for the hydrogen and oxygen sides. The cell was connected to two separate gas collection cylinders (shown in Figure 9) each, for oxygen and hydrogen gas.

The experimental setup of the modified URFC or proton battery is shown in Figure 10 along with schematic diagram in Figure 11. The oxygen side of the proton battery had three outlets, two for oxygen and one for water intake. On the hydrogen side, there were two vents for extraction of hydrogen gas. In the process of electrolysis, the water inlet was connected to the lower vent of the anode side end plate, oxygen was collected from the upper vent, and the third vent was blocked. On the hydrogen side, the upper vent was used as hydrogen outlet, whereas the bottom vent was blocked. A variable 30 V DC source was used to supply electric charge in the case of electrolyzer mode of the modified URFC. Two multimeters were used to measure the voltage across and current through the cell.

Processes 2019, *7*, 238

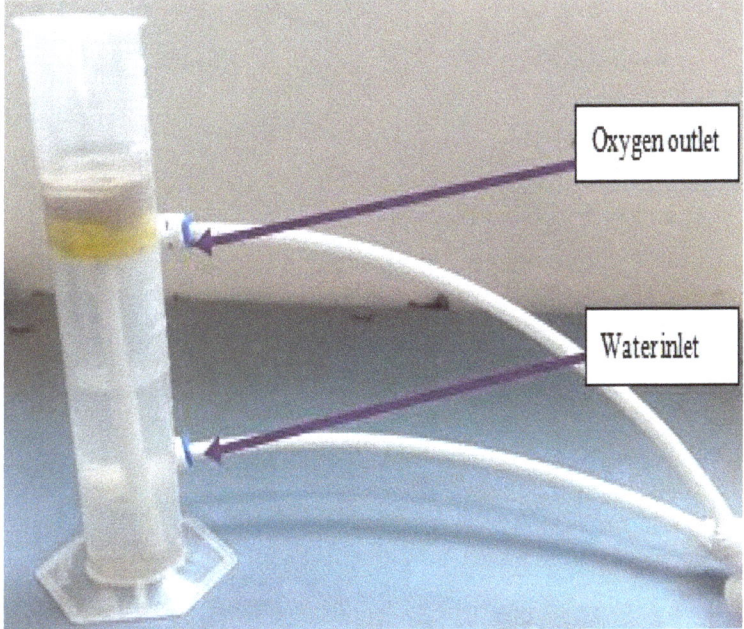

Figure 9. Oxygen side gas collection cylinder of the modified unitized regenerative fuel cell URFC.

Figure 10. Experimental setup of proton battery or modified unitized regenerative fuel cell URFC.

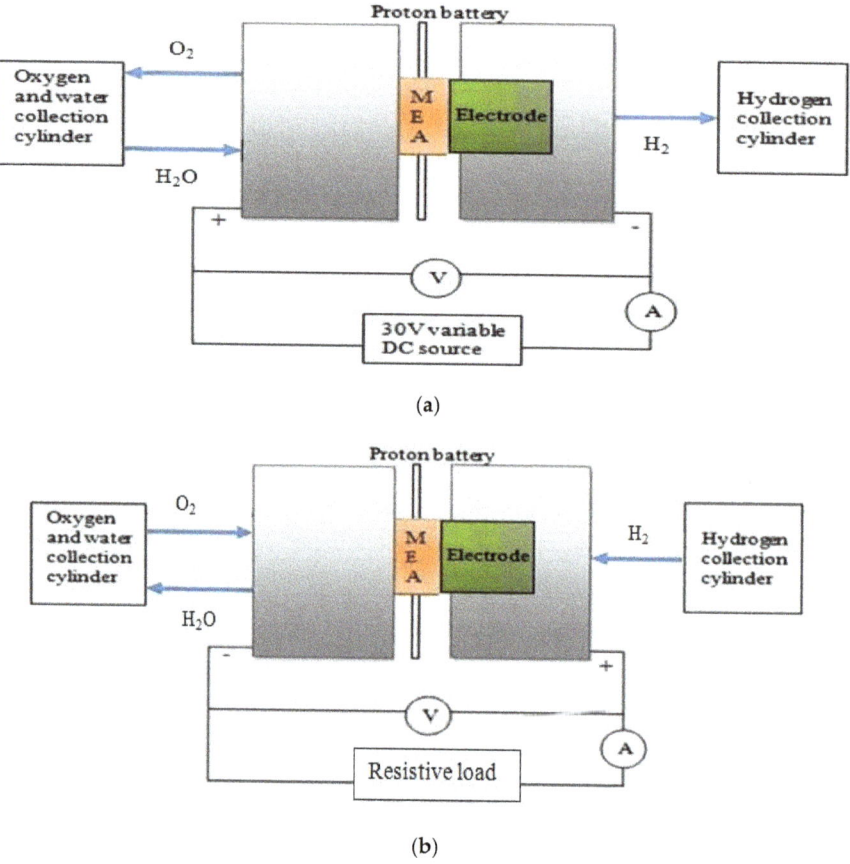

Figure 11. (a) Schematic diagram of charging mode of the modified unitized regenerative fuel cell (URFC) or proton battery; (b) schematic diagram of discharging mode of modified URFC or proton battery.

For galvanostatic charging, the potential across the battery was initially kept at 1.5 V and increased by 0.1 V after every half hour. With the rise in potential, the current across the battery increased. The potential of battery was raised till rapid evolution of hydrogen gas was observed in the gas collection cylinder indicating that the storage was full. The operation was stopped thereafter. Water supplied was dissociated into oxygen, H^+, and electron under the influence of electric potential and catalyst. The produced oxygen gas was allowed to move out of the cell and was collected, whereas H^+ passed through the polymer membrane (Nafion 117 in this case) and electron travelled through the electrical circuit. Hydrogen ions emerging on the polymer membrane at cathode get stabilized with electron before entering into the porous storage electrode (MWCNT electrode in this case). Hydrogen atoms get adsorbed in the MWCNT electrode either by forming a weak chemical bond on the internal surface of pores (called chemisorption) or by getting physically adsorbed inside the tiniest pores (ultramicropores) due to Van der Waals forces.

The experimental set up was made to rest for one hour after charging process for discharging. While discharging, the cell was connected to an electrical load in order to start drawing current out of it. Under the influence of potential difference, the weak surface chemical bond breaks up and the hydrogen atom comes out of the storage. Here hydrogen reduction (HRR) reaction takes place, in which the platinum black catalyst breaks the hydrogen atom into H^+ and electron. Hydrogen ion and electron travels back towards the oxygen side through the membrane and the electrical circuit, respectively; where they react with the oxygen produced while charging to reform water. The weight of the produced oxygen and hydrogen and the charging and discharging capacity in mAh/g were recorded and used later in calculation to ascertain the electrochemical hydrogen storage capacity of the fabricated MWCNT electrode.

3. Results

During galvanostatic charging, the voltage was increased by 0.1 V in an interval of thirty minutes. From Figure 12, it can be inferred that current increased with voltage. The maximum rise in current observed was 230 mA. At 2.6 V, the charging process was stopped due to the sudden increment in production of hydrogen gas, indicating that the storage was full (i.e., the hydrogen ions that were emerging on the membrane were combining with electrons to form hydrogen gas, instead of entering into the storage electrode). The sharp dip in the curve, as shown in Figure 12, was due to the electric potential predominantly being utilized for water disassociation. Once the electric potential reached a value that was enough to break the inter-molecular forces associated with hydrogen and oxygen in water, the current started rising (refer to Figure 12 and Table 7).

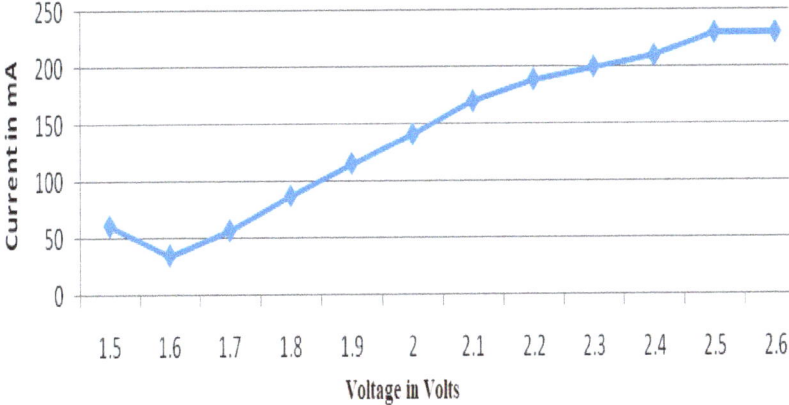

Figure 12. Graph representing relation between current and voltage while charging or during electrolyzer mode of proton battery.

Table 7. X and Y axis variable values of the active points shown in Figure 12.

Current (mA)	Voltage (V)
60.5	1.5
34.5	1.6
56.5	1.7
86.7	1.8
115.4	1.9
140.8	2
169.8	2.1
188.6	2.2
200	2.3
210	2.4
230	2.5
230	2.6

Voltage was increased after specified time intervals that lead to an increase in current and, hence, the production of hydrogen ions. For the first 1.5 h of charging, no hydrogen gas generation was observed in the collection cylinder. However, the cut-in voltage (i.e., where current starts rising after the dip shown in Figure 12) of 1.5 V resulted in generation of oxygen gas, which was observed in the form of bubbles in the collection cylinder (shown in Figure 9). It was deemed that the corresponding hydrogen got absorbed in the electrode and there was no visible sign of hydrogen generation. Figure 13 represents the hydrogen generation rate during charging of the cell. It is clearly visible in the figure that the cut-in voltage was reached at 1.5 V in 1.5 h, after which the dissociated hydrogen moved out of the cell and bubbled through water in the collection cylinder. Although the presented work aims to suppress the formation of hydrogen gas, practically not all the emerging H^+ ions on the electrolytic membrane get adsorbed in the storage (i.e., certain H^+ ions do combine with e^- and liberate as H_2). A fairly linear behavior of the hydrogen gas generation rate (i.e., the slope of curve in Figure 13) was observed with respect to increase of the applied voltage and corresponding current, as is clear from Figure 13 and Table 8.

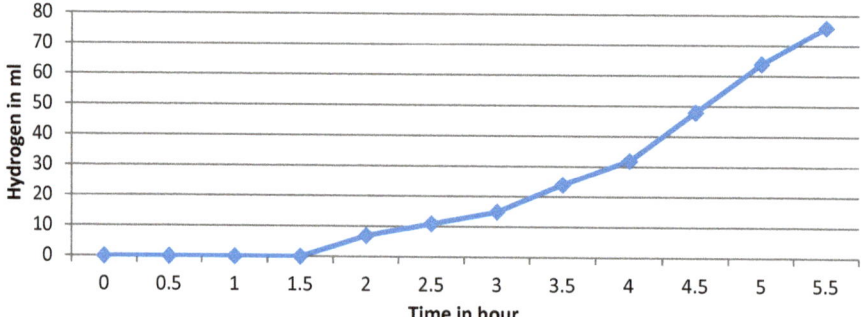

Figure 13. Graph representing relation between hydrogen gas produced and time during charging of proton battery.

Table 8. X and Y axis variable values of the active points shown in Figure 13.

Hydrogen (mL)	Time (h)
0	0
0	0.5
0	1
0	1.5
7	2
11	2.5
15	3
24	3.5
32	4
48	4.5
64	5
76	5.5

The production of oxygen started when water was subjected to cut-in voltage of 1.5 V, as shown in Figure 14 and Table 9. The oxygen produced was stored in the gas collection cylinder and its production increased with the rise in the voltage. The higher the applied voltage, the higher would be the charge flow through the circuit, leading to the increase in the production of oxygen ions that combine to form oxygen gas and release electrons in the presence of the catalyst. This was verified through Faraday's law given in Equation (1).

$$M_o = \frac{I \times t}{4F \times 1000} \times \frac{R \times T}{P}. \tag{1}$$

where,

Mo→ mass of oxygen generated in Kg.
I→ current in amperes.
t→ time in seconds.
F→ Faraday's constant = 96,485 J per volt gram equivalent.
R→ gas constant = 8.314 J/mol·K.
T→ ambient temperature in Kelvin.
P→ atmospheric pressure in kilo Pascals.

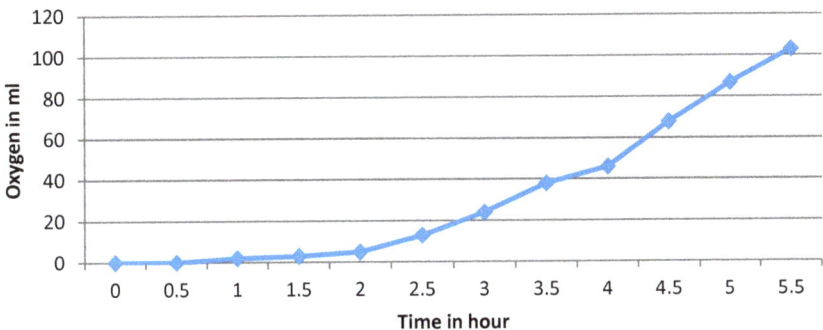

Figure 14. Graph showing relation between oxygen gas produced with respect to time in case of charging of proton battery.

Table 9. X and Y axis variable values of the active points shown in Figure 14.

Oxygen (mL)	Time (h)
0	0
0	0.5
2	1
3	1.5
5	2
13	2.5
24	3
38	3.5
46	4
68	4.5
87	5
103	5.5

During electrolyzer mode or charging, the amount of hydrogen produced was calculated by Faraday's law given in Equation (2).

$$M = \frac{I \times t}{F \times 1000}. \qquad (2)$$

where,

M→ theoretical mass of hydrogen generated in kg.
I→ discharge current in mA.
t→ discharge time in sec.
F→ Faraday's constant (96,485 C mol^{-1}).

Mass of the produced hydrogen was subtracted from theoretically calculated mass of hydrogen to ascertain the amount of hydrogen adsorbed in the porous MWCNT electrode. After calculating the mass of hydrogen stored in the electrode, the percent of hydrogen stored in MWCNT was calculated by Equation (3).

$$H_c\% = \frac{H}{H+C} \quad (3)$$

where,

$H_C \rightarrow$ weight percent of hydrogen stored in MWCNT.
$H \rightarrow$ mass of hydrogen stored in MWCNT.
$C \rightarrow$ mass of MWCNT used in electrode.

The weight percent of hydrogen was calculated for successive time intervals of 30 min, and its variation with respect to the applied current is shown in Figure 15 and Table 10. The initial plateau in the curve signifies the amount of electric potential that got invested in overcoming the inter-molecular forces of H_2O and; therefore, lead to lower hydrogen storage, as shown in Figure 15. Once the water started disassociating, a linear increase in hydrogen storage with respect to the applied voltage was observed. Another plateau between 1.8 and 2.1 V was due to the saturation of the storage space and, finally, after 2.1 V, a clear drop in the curve could be seen that signifies that storage is almost full, thereby increasing the H_2 generation rate rapidly.

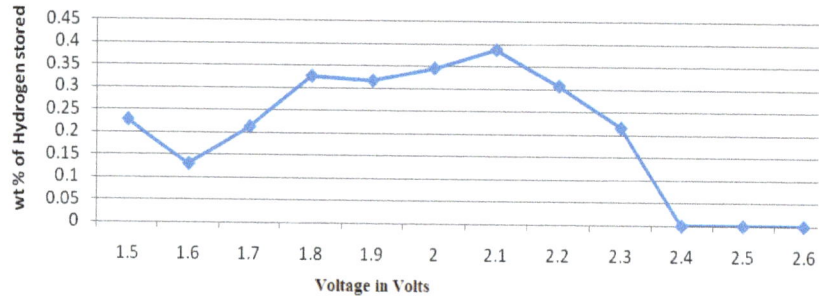

Figure 15. Graph showing relation between wt% of hydrogen stored in carbon and voltage during charging of modified URFC or proton battery.

Table 10. X and Y axis variable values of the active points shown in Figure 15.

Hydrogen Stored (wt%)	Voltage (V)
0.228056794	1.5
0.130176499	1.6
0.213010777	1.7
0.326496125	1.8
0.317086481	1.9
0.345488992	2
0.38737332	2.1
0.307371183	2.2
0.216174703	2.3
0	2.4
0	2.5
0	2.6

Figure 16 and Table 11 are an outcome of the curves presented in Figures 12 and 15 for verification of the obtained results. It represents the wt% of hydrogen getting adsorbed electrochemically in the porous MWCNT electrode with respect to current. From Figure 16, it can be suggested that for electrochemical hydrogen storage in porous storage material, the charging current has to be maintained at a low value to discourage generation of hydrogen gas.

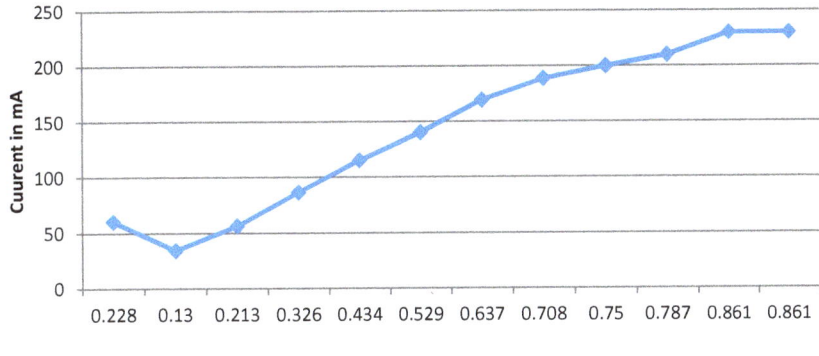

Figure 16. Graph showing relation between wt% of hydrogen stored in carbon and current during charging of modified unitized regenerative fuel cell (URFC) or proton battery.

Table 11. X and Y axis variable values of the active points shown in Figure 16.

Current (mA)	Hydrogen Stored in Carbon (wt%)
60.5	0.228056794
34.5	0.130176499
56.5	0.213010777
86.7	0.326496125
115.4	0.317086481
140.8	0.345488992
169.8	0.38737332
188.6	0.307371183
200	0.216174703
210	0
230	0
230	0

The graph between hydrogen wt% stored in MWCNT with respect to time is shown in Figure 17 and corresponding values are presented in Table 12. The maximum percent of hydrogen stored in carbon was 0.387% after 3 h of electrolysis. The possible reason for decay in hydrogen storage in MWCNT is that its pores get filled after a certain time. Since the voltage was applied at time (t) = 0 and the applied voltage was increased with respect to time, that is why the shape of curves in both Figures 15 and 17 were the same, which, thereby, justifies the authenticity of the obtained results.

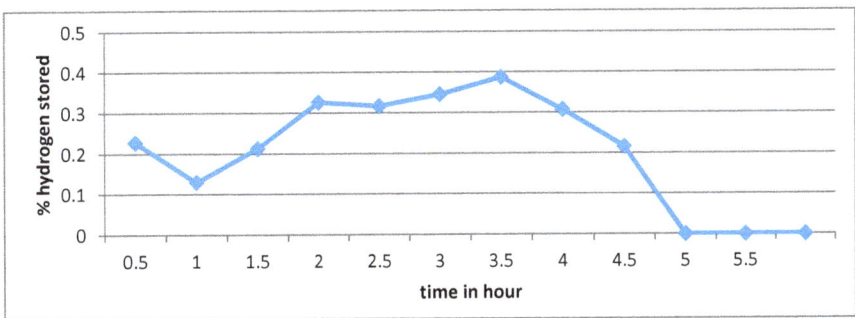

Figure 17. Graph showing the relation between wt% of hydrogen stored in carbon and time during charging of modified unitized regenerative fuel cell (URFC) or proton battery.

Table 12. X and Y axis variable values of the active points shown in Figure 17.

Hydrogen Stored in Carbon (wt%)	Time(h)
0.228056794	0
0.130176499	0.5
0.213010777	1
0.326496125	1.5
0.317086481	2
0.345488992	2.5
0.38737332	3
0.307371183	3.5
0.216174703	4
0	4.5
0	5
0	5.5

The consolidated results obtained during a single cycle of charging and discharging of the fabricated modified URFC are presented in Table 13 and compared with those reported in literature [14].

Table 13. Comparison among the obtained results and those reported in literature.

Name of the Sample	Equivalent Weight of Hydrogen Stored in Carbon during Charging (wt%)	Equivalent Weight of Hydrogen Stored in Carbon during Discharging (wt%)	Equivalent Weight of Hydrogen Stored in Carbon during Charging (mAh/g)	Equivalent Weight of Hydrogen Stored in Carbon during Discharging (mAh/g)
MWCNT	2.47	2.45	666.9	661.5
Activated Carbon [14]	1.0	0.8	270	216

The SEM test was performed on the oxygen side CC after the experiment, as shown in Figure 18. It was found that stands of carbon cloth got broken during the experimental procedure due to oxidation.

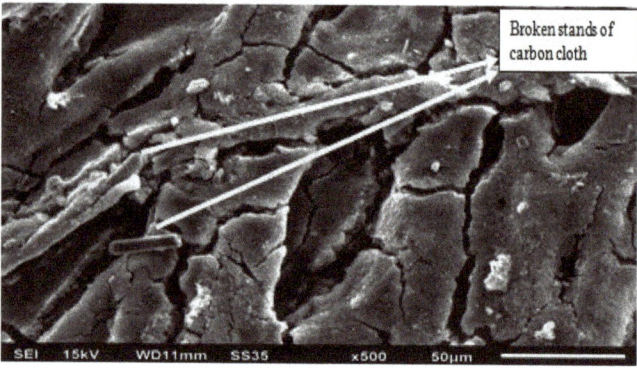

Figure 18. Scanning Electron Microscopy (SEM) image of the oxygen side carbon cloth used as a gas diffusion layer.

4. Discussion

As per the chemical composition of water, it was deemed that, in dissociation, two hydrogen atoms would release. Ideally, the amount of hydrogen produced should be double the amount of oxygen, but it was observed that the amount of hydrogen produced was far less than the produced oxygen; presuming no leakage and negligible loss, the deficit hydrogen got adsorbed electrochemically in the porous MWCNT electrode. The produced hydrogen got adsorbed physically in the porous MWCNT electrode, simultaneously allowing few atoms to form hydrogen gas. The adsorption process [25] was in two forms—chemisorption and physisorption. During chemisorption, there was a weak chemical bond formation between the host material, hydrogen ion, and electron, whereas in the case of physisorption, weak Van der Waals forces between the surface of the host and hydrogen were formed. Whereas during desorption, electrical load was applied across the cell. The charging capacity of the cell was found to be 666.9 mAh/g, which is equivalent to 2.47 wt%, and is significantly higher than reported in literature published in 2018 [15]. The discharging capacity of the cell was found to be 661.5 mAh/g, which is equivalent to 2.45 wt%, and higher than the reported literature of 2018 [15]. US DOE target of hydrogen in terms of wt% is 4.0 wt% for single use and 3.0% for rechargeable [26]. However, this target is for gaseous storage in various materials, whereas in the presented work the hydrogen is being stored in solid-state (ionic form-H^+), which is the safest form compared to other storage forms. In the presented work, generation of hydrogen gas is suppressed by operating at low voltage and current ranges. Additionally, the gaseous hydrogen storage needs the pressure of 700 bar and in the presented work the operating pressure is atmospheric (i.e., 1 bar). Therefore, we cannot compare the achieved gravimetric energy densities with US DOE targets for gaseous hydrogen storage.

5. Conclusions

It is experimentally demonstrated that hydrogen can be stored in solid-state in a MWCNT electrode integrated in a modified URFC or proton battery. It is evident from the obtained results that the hydronium ions (H_3O^+) travelled through the MEA, from anode to cathode, through the polymer electrolyte and got adsorbed in the porous electrode. The employed cell integrated with the porous MWCNT electrode is charged with capacity of 666.9 mAh/g, which is equivalent to 2.47 wt%, whereas the discharged capacity is 661.5 mAh/g, which is equivalent to 2.45 wt% and is comparable with the U.S. DOE targets for hydrogen storage. However, rapid cyclic charge–discharge testing is required to be carried out in future to ascertain the life cycle of the proton battery. It is also observed, in experimentation, that the carbon cloth on the anode side, which served as a gas diffusion layer (GDL), got oxidized during the charging mode. Therefore, for GDL, it is recommended to use alternate material, like porous titanium felt/frit. The obtained result proves the technical feasibility of a modified URFC with an integrated MWCNT-based hydrogen storage electrode. However, other measures are to be taken to enhance the gravimetric energy storage density, like the usage of alternate proton conducting medium within the porous hydrogen storage electrode.

Author Contributions: All the listed authors have contributed significantly in the present research work and the individual contributions are articulated below. Design of the experiments and first draft writing, D.K.; Fabrication of electrode and gas collection cylinder, D.K.; A.S.O.; Design of the electrical circuits, validation and formal analysis, P.N.; Supervision, writing review and editing, A.S.O. and P.N.

Funding: This research received no external funding.

Conflicts of Interest: The authors declare no conflicts of interest.

References

1. Ecology Webinar Series. Available online: http://www.ecology.com/2011/09/03/the-history-of-energy-use/ (accessed on 6 February 2019).
2. Riahi, K.; Rao, S.; Krey, V.; Cho, C.; Chirkov, V.; Fischer, G.; Kindermann, G.; Nakicenovic, N.; Rafaj, P. RCP 8.5—A scenario of comparatively high greenhouse gas emissions. *Clim. Chang.* **2011**, *109*, 33. [CrossRef]
3. Dufour, J.; Serrano, D.P.; Galvez, J.L.; Moreno, J.; Garcia, C. Life cycle assessment of processes for hydrogen production. *Environmental feasibility and reduction of greenhouse gases emissions. Int. J. Hydrog. Energy* **2009**, *34*, 1370–1376. [CrossRef]
4. Edwards, P.P.; Kuznetsov, V.L.; David, W.I.F.; Brandon, N.P. Hydrogen and fuel cells: Towards a sustainable energy future. *Energy Policy* **2008**, *36*, 4356–4362. [CrossRef]
5. Ali, M.; Ekström, J.; Lehtonen, M. Sizing hydrogen energy storage in consideration of demand response in highly renewable generation power systems. *Energies* **2018**, *11*, 1113. [CrossRef]

6. Rosen, M. Energy sustainability: A pragmatic approach and illustrations. *Sustainability* **2009**, *1*, 55–80. [CrossRef]
7. Gracia, L.; Casero, P.; Bourasseau, C.; Chabert, A. Use of Hydrogen in Off-Grid Locations, a Techno-Economic Assessment. *Energ.* **2018**, *11*, 3141. [CrossRef]
8. Wang, F.C.; Hsiao, Y.S.; Yang, Y.Z. The Optimization of Hybrid Power Systems with Renewable Energy and Hydrogen Generation. *Energies* **2018**, *11*, 1948. [CrossRef]
9. Zhang, F.; Cooke, P. Hydrogen and fuel cell development in China: A review. *Eur. Plan. Stud.* **2010**, *18*, 1153–1168. [CrossRef]
10. Züttel, A.; Borgschulte, A.; Schlapbach, L. *Hydrogen as a Future Energy Carrier*; Wiley Publications: Hoboken, NJ, USA, 2008; ISBN 9783527622894.
11. Shabani, B.; Andrews, J. *Hydrogen and Fuel Cells*; Energy Sustainability through Green Energy; Springer: New Delhi, Indai, 2015; ISBN 9788132223375.
12. Mohammadi, S.S. Investigation of a Reversible PEM Fuel Cell with Integrated Metal-Hydride Hydrogen Storage. Master's Thesis, RMIT, Melbourne City Campus, Victoria, Australia, 2013.
13. Xu, Y.; He, G.; Wang, X. Hydrogen evolution reaction on the AB_5 metal hydride electrode. *Int. J. Hydrog. Energy* **2003**, *28*, 961–965. [CrossRef]
14. Jazaeri, M.J. The Feasibility of a Unitised Regenerative Fuel Cell with a Reversible Carbon-Based Hydrogen Storage Electrode. Master's Thesis, RMIT, Melbourne City Campus, Victoria, Australia, 2013.
15. Heidari, S.; Mohammadi, S.S.; Oberoi, A.S.; Andrews, J. Technical feasibility of a proton battery with an activated carbon electrode. *Int. J. Hydrog. Energy* **2018**, *43*, 6197–6209. [CrossRef]
16. Liu, E.; Wang, J.; Li, J.; Shi, C.; He, C.; Du, X.; Zhao, N. Enhanced electrochemical hydrogen storage capacity of multi-walled carbon nanotubes by TiO_2 decoration. *Int. J. Hydrog. Energy* **2011**, *36*, 6739–6743. [CrossRef]
17. Mosquera, E.; Diaz-Droguett, D.E.; Carvajal, N.; Roble, M.; Morel, M.; Espinoza, R. Characterization and hydrogen storage in multi-walled carbon nanotubes grown by aerosol-assisted CVD method. *Diam. Relat. Mater.* **2014**, *43*, 66–71. [CrossRef]
18. Erünal, E.; Ulusal, F.; Aslan, M.Y.; Güzel, B.; Üner, D. Enhancement of hydrogen storage capacity of multi-walled carbon nanotubes with palladium doping prepared through supercritical CO_2 deposition method. *Int. J. Hydrog. Energy* **2018**, *43*, 10755–11076. [CrossRef]
19. Reyhani, A.; Mortazavi, S.Z.; Mirershadi, S.; Moshfegh, A.Z.; Parvin, P.; Golikand, A.N. Hydrogen storage in decorated multiwalled carbon nanotubes by Ca, Co, Fe, Ni, and Pd nanoparticles under ambient conditions. *J. Phys. Chem. C* **2011**, *115*, 6994–7001. [CrossRef]
20. Khoshnevisan, B.; Behpour, M.; Ghoreishi, S.M.; Hemmati, M. Absorptions of hydrogen in Ag–CNTs electrode. *Int. J. Hydrog. Energy* **2007**, *32*, 3860–3863. [CrossRef]
21. Nangsuay, A.; Ruangpanit, Y.; Meijerhof, R.; Attamangkune, S. Yolk absorption and embryo development of small and large eggs originating from young and old breeder hens. *Poult. Sci.* **2011**, *90*, 2648–2655. [CrossRef] [PubMed]
22. Achaw, O.W. A study of the porosity of activated carbons using the scanning electron microscope. *Scanning Electron. Microsc.* **2012**. [CrossRef]
23. Béguin, F.; Kierzek, K.; Friebe, M.; Jankowska, A.; Machnikowski, J.; Jurewicz, K.; Frackowiak, E. Effect of various porous nanotextures on the reversible electrochemical sorption of hydrogen in activated carbons. *Electrochim. Acta* **2006**, *51*, 2161–2167. [CrossRef]
24. Turan, C.; Cora, Ö.N.; Koç, M. Effect of manufacturing processes on contact resistance characteristics of metallic bipolar plates in PEM fuel cells. *Int. J. Hydrog. Energy* **2011**, *36*, 12370–12380. [CrossRef]
25. Babel, K.; Janasiak, D.; Jurewicz, K. Electrochemical hydrogen storage in activated carbons with different pore structures derived from certain lignocellulose materials. *Carbon* **2012**, *50*, 5017–5026. [CrossRef]
26. DOE Technical Targets for Hydrogen Storage Systems for Portable Power Equipment. Available online: https://www.energy.gov/eere/fuelcells/doe-technical-targets-hydrogen-storage-systems-portable-power-equipment (accessed on 21 March 2019).

© 2019 by the authors. Licensee MDPI, Basel, Switzerland. This article is an open access article distributed under the terms and conditions of the Creative Commons Attribution (CC BY) license (http://creativecommons.org/licenses/by/4.0/).

Article

Enhanced Lifetime Cathode for Alkaline Electrolysis Using Standard Commercial Titanium Nitride Coatings

William J. F. Gannon, Daniel R. Jones and Charles W. Dunnill *

Energy Safety Research Institute, Swansea University Bay Campus, Fabian way, Swansea SA1 8EN, UK; 920920@Swansea.ac.uk (W.J.F.G.); d.r.jones@swansea.ac.uk (D.R.J.)
* Correspondence: c.dunnill@swansea.ac.uk; Tel.: +44-179-266-6230

Received: 10 January 2019; Accepted: 18 February 2019; Published: 21 February 2019

Abstract: The use of hydrogen gas as a means of decoupling supply from demand is crucial for the transition to carbon-neutral energy sources and a greener, more distributed energy landscape. This work shows how simple commercially available titanium nitride coatings can be used to extend the lifetime of 316 grade stainless-steel electrodes for use as the cathode in an alkaline electrolysis cell. The material was subjected to accelerated ageing, with the specific aim of assessing the coating's suitability for use with intermittent renewable energy sources. Over 2000 cycles lasting 5.5 days, an electrolytic cell featuring the coating outperformed a control cell by 250 mV, and a reduction of overpotential at the cathode of 400 mV was observed. This work also confirms that the coating is solely suitable for cathodic use and presents an analysis of the surface changes that occur if it is used anodically.

Keywords: titanium nitride; stainless steel; alkaline electrolysis; energy storage

1. Introduction

Developing high-performance electrode coatings for water splitting under room temperature, alkaline conditions remains of paramount importance as a means of storing excess renewable energy as hydrogen gas [1,2]. The use of hydrogen gas to decouple supply from demand is crucial for the transition to intermittent supplies of renewable energy and a greener energy landscape [3,4]. Alkaline electrolysis provides an alternative to Proton Exchange Membrane (PEM) electrolysis without the high costs, whilst retaining the high efficiency [5,6]. Storing energy cheaply and efficiently by conversion into hydrogen allows for a sustainable "Many-to-Many" energy landscape whereby multiple small-scale distributed energy inputs feed into the system, compared to the now outdated "One-to-Many" system whereby a single power station feeds to many houses and businesses. The "Many-to-Many" approach is far more resilient and sustainable given the intermittent nature of green energy inputs.

This paper examines the electrode stability and lifetime potential for stainless steel and a commercially coated titanium nitride electrode setup. The composition of stainless steels is governed by international standards and they are widely available, making them a cost-effective source of a reliable substrate material. Certainly, they are more widely available and cost-effective than a pure nickel substrate. The most common grades investigated for electrolysis are 304 (304SS) and 316 (316SS) [7–9], both of which are composed primarily of Fe, Ni and Cr, but with 316SS featuring a higher percentage of Ni, in addition to about 2.5 wt % Mo. This affords 316SS greater resistance to corrosion, making it the first choice for marine and medical applications, but also making it more expensive, and potentially more difficult to obtain coating adherence. From the results reported by Carta et al. an overpotential for a 316SS cathode of −0.34 V at 10 mA cm^{-2} was observed [8], but it has been clear to the authors over extended use that even 316SS will experience significant cathodic corrosion (cf. Figure 7).

Titanium nitride is known for its high thermal and electrical conductivity, in addition to its mechanical hardness [10], and is known to enhance electrode lifetime for the oxygen and hydrogen reduction reactions in a proton exchange membrane fuel cell [11]. However, to our knowledge it has not been investigated for use with intermittent alkaline electrolysis, so this work is the first study of its kind. In addition, it is widely available as a bespoke coating service for the lifetime enhancement of machine tools, alongside alternative coatings such as CrN, TiAlN and WS_2 [12]. In this context, the coatings are selected for their extreme hardness, and their ability to resist oxidation at the high temperatures generated during machining (which can exceed 800 °C). These properties are of secondary concern for room-temperature alkaline electrolysis, but the wide availability and accessibility of the coatings makes them potentially cost-effective. However, it remains to be seen which (if any) of these coatings constitutes the optimal trade-off between performance and cost. This study will focus on titanium nitride.

Accelerated Ageing

Electrode lifetime is often studied and reported within the scientific literature by employing constant currents, which are arguably of limited applicability in the field of intermittent renewable energy capture. Intermittent usage is recognised to play a significant role in the breakdown of electrocatalysts, especially ones involving nickel [13]. Therefore, to achieve rapid ageing, it was desired not only to subject the electrodes to large total quantities of current, but also to the destructive on-off cycling caused by the intermittent nature of renewable energy. A regime was devised that consisted of 2 min on, followed by 2 min under open-circuit conditions, permitting the electrodes enough time to both fully charge and discharge within each cycle, applying the corrosive stress on each change of phase. Room-temperature electrolysis is chosen because it is more applicable to renewable energy applications, where the demands of intermittency mean that permanent heating of the electrolyte is unsustainable.

2. Materials and Methods

2.1. Electrodes

The electrodes used were all 316-grade stainless steel 0.9 mm thickness cut to size and shape, either as supplied or coated in TiN using a standard commercial preparation delivering a 1 to 4 µm thick TiN coating. The commercial titanium nitride coatings were applied by Wallwork Cambridge Ltd., UK, and their internal specification for the coating was "TIN COAT SPEC 300 1–4 MICRONS". In order to respect the intellectual property of the company, no attempts have been made to reverse-engineer the coatings, so it is not possible to report in this study how the coating thickness might affect performance. It is not expected that differences in coating thickness above 1 µm will cause significant changes. This is because the material is electrically conductive, and because such a thickness constitutes many thousands of atoms, it is unlikely to affect the surface chemistry.

2.2. Ageing

All ageing experiments were conducted using a constant current power supply, connected to a two-electrode laminated electrolytic cell, which was primarily comprised of laser-cut acrylic plastic (see Figure 1) [14]. A Zirfon™ membrane was used to keep the evolved gases separate, and the distance between electrodes was approximately 30 mm [5]. Two different ageing protocols were followed, as detailed in Table 1. It was observed that at the higher current density a stronger electrolyte was needed to keep the total voltage drop across the electrolytic cell, and with it the associated ohmic heating, within reasonable limits. The choice to use NaOH instead of the more usual KOH was made because it is cheaper, and therefore more practical for commercial applications. Also, even though the safety data sheet states that it must never be disposed of down the drain, it is widely recognized and used as a drain cleaner, and therefore accidental spillage or leakage into the sewer is a manageable environmental hazard.

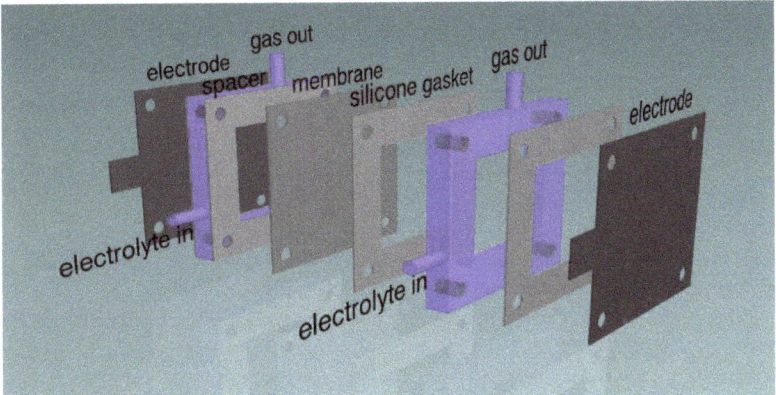

Figure 1. Typical design of experimental electrolytic cell, based on laser-cut components.

Table 1. Different accelerated ageing protocols employed.

Ageing Protocol	Current (mA cm^{-2})	Cycles	Electrolyte
A	100	2000 to 2500	0.5 M NaOH
B	200	2000 to 2500	1 M NaOH

Each experiment lasted between 5.5 and seven days. made up of 4-min cycles. In some experiments the electrolyte was circulated through a single external chamber using a small brushless pump.

2.3. Three Electrode Experiments

All three-electrode experiments were conducted on an Ivium (Eindhoven, The Netherlands) n-Stat potentiostat, connected to a similar design of electrolytic cell. The working electrode (WE) surface area was reduced to 9 cm^2 by gluing or bolting the electrode to a laser cut sheet of 3 mm clear acrylic containing a 3 × 3 cm window. The counter electrode was a 316 stainless-steel plate (of which 6 × 6 cm was exposed), and the reference electrode (RE) was a commercial design involving a Ag/AgCl wire suspended in 3 M KCl. The distance between working and counter electrodes was approximately 15 mm. The electrolyte was 0.5 M NaOH (standard reagent grade) and deionised water was used throughout. Before each experiment the RE was checked against a standard calomel electrode (SCE), and the electrolyte was bubbled with nitrogen for 10 min to reduce dissolved oxygen. All experiments were conducted at laboratory ambient temperature, which was 20 ± 1 °C.

2.4. Tafel Slope

The procedure outlined by Stevens et al. was followed to obtain measurements of the Tafel slope [15]. This involved chronopotentiometry steps at varying current densities and durations as specified in Table 2.

Each experimental run consisted of both ascending and descending Tafel slopes, with the whole procedure repeated twice. The published results are taken from the descending slope of the second run.

Table 2. Current densities and durations for Tafel analysis.

Step	Current Density (A cm^{-2})	Duration (Seconds)
1,14	1.0×10^{-5}	480
2,13	3.2×10^{-5}	480
3,11	1.0×10^{-4}	240
4,10	3.2×10^{-4}	120
5,9	1.0×10^{-3}	120
6,8	3.2×10^{-3}	120
7	1.0×10^{-2}	120

2.5. iR Correction

In order to correct for voltage losses in the electrolyte between the reference and working electrodes, electro impedance spectroscopy (EIS) was performed between 100 Hz and 1 MHz. The series resistance of the electrolyte was then defined as the magnitude of the point of closest approach to the origin of the resulting Nyquist plot. The voltage drop across the electrolyte could then be cancelled out simply by multiplying this resistance by the total cell current.

2.6. Electron Microscope

Scanning electron microscope (SEM) imaging and energy-dispersive spectroscopy (EDX) were performed on an Oxford Instruments (Oxford, UK) AZtecOne spectrometer attached to a Hitachi (Tokyo, Japan) TM3030 table top microscope.

3. Results and Discussion

Cyclic voltammetry experiments revealed that the TiN coating cannot be employed as an anode, even for brief periods, which is in agreement with previous findings [16,17]. The positive electric potential caused the coating to change rapidly from the original gold colour to orange-brown in just a few tens of seconds, accompanied by the loss of the great majority of the electrical conductivity. The physical origin of this change was investigated using XPS, as described in Section 3.2. Any TiN-coated electrodes present in a commercial electrolyser would therefore be at significant risk of destruction should the incorrect polarity mistakenly be applied.

The electrical performance comparison of both the stainless-steel and TiN-coated cathodes is as shown in Figure 2. These results show that the Tafel slopes for both materials are very nearly the same. Both are largely parallel to the 120 mV/decade value (shown as a dashed line), which is the value anticipated from theoretical calculations within the literature [18]. However, the coated electrode requires approximately an extra 300 mV of overpotential to achieve the same current density as the uncoated material.

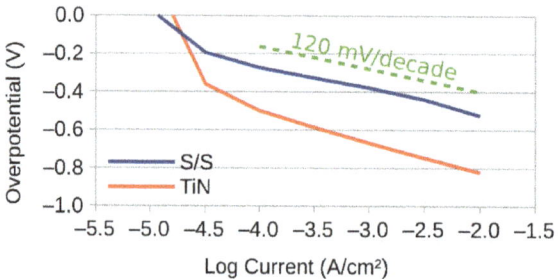

Figure 2. Tafel plots for stainless-steel and TiN before ageing.

Figure 3 shows the visual representation of the electrodes after Ageing Protocol A. Here it can be seen that the TiN-coated cathode (Figure 3e) retained much of its original golden colour, as opposed to the uncoated stainless steel (Figure 3c), which became almost purple.

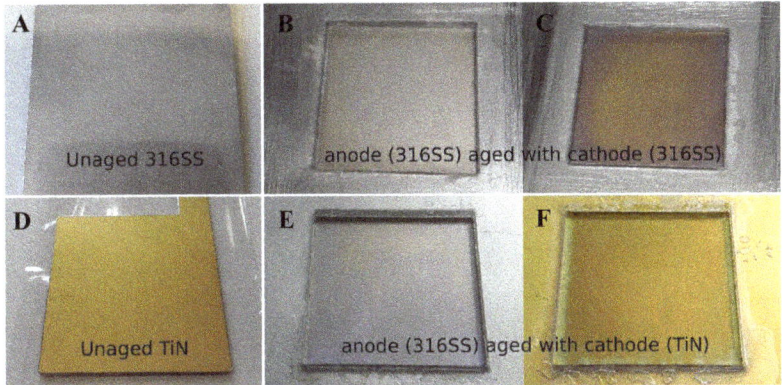

Figure 3. Visual appearance before and after Ageing Protocol A of 316 stainless-steel electrodes both coated in TiN and uncoated. (**A**) is original unaged electrode, (**B**) and (**C**) are the anode and cathode respectively), (**D**) is the unaged TiN coated electrode, (**E**) and (**F**) are the anode and cathode respectively.

During these experiments, neither combination exhibited much variation in two-electrode electrical performance, as shown in Figure 4. This indicates that both materials are electrically stable long term, at least at this current density. Note also that the TiN cell consistently required approximately an extra 250 mV, which agrees well with the three-electrode results presented in Figure 2. We may therefore conclude that most of this additional voltage is a result of the TiN coating on the cathode.

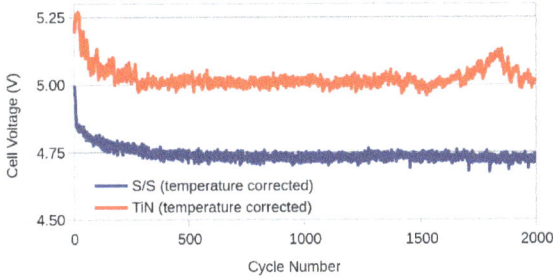

Figure 4. Variation of cell voltage over 2000 cycles of Ageing Protocol A.

Further long-term tests were conducted at 200 mA cm^{-2} to test whether the electrical performance of the uncoated stainless-steel cathode would degenerate if subjected to greater accelerated ageing, as shown in Figure 5.

Here it can be seen that the voltage was lower in general, due to the higher concentration of the electrolyte. Also, the initial 300 mV difference was quickly overturned, and thereafter the gap gradually extended until finally the TiN cell outperformed the stainless-steel cell by about 250 mV. This performance improvement is not necessarily attributable to the cathode, since it is a two-electrode cell, and therefore does not permit an individual assessment to be made of either electrode. To assess this, three-electrode experiments were again performed to measure the electrical performance of the cathodes in isolation, as shown in Figure 6.

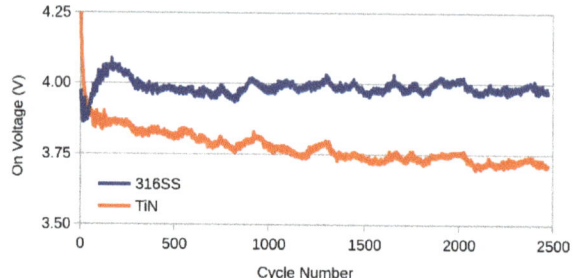

Figure 5. Variation of cell voltage over 2000 cycles of Ageing Protocol B.

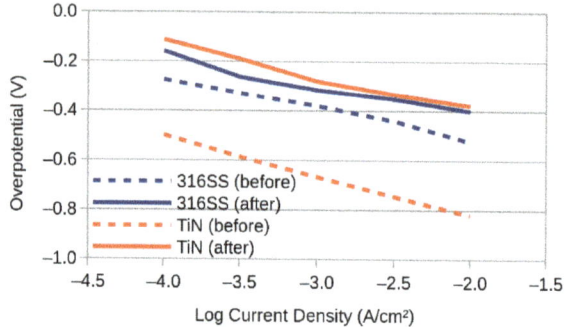

Figure 6. Tafel plots before and after Ageing Protocol B.

The TiN cathode was now able to outperform the stainless cathode, although only by about 20 mV. The improvement seen in Figure 5. is therefore likely due to factors other than the cathode alone, such as the electrolyte, membrane or anode. Perhaps of greatest significance is the observation that the anode used in conjunction with the TiN cathode had taken on a coppery appearance, and it is possible that this has led to an unexpected and unexplained increase in performance. Also significant is that both cathodes improved over the course of the experiment, with the TiN cathode overpotential decreasing by a remarkable 400 mV.

Despite this improvement in electrical performance, both electrodes exhibited significant deterioration in their visual appearance, as shown in Figure 7. Nevertheless, it is perhaps indicative of their potential for real-world longevity that their appearance improved after immersion for three days in fresh 0.5 M NaOH (see right-hand images of Figure 7), despite being almost completely black immediately after accelerated ageing.

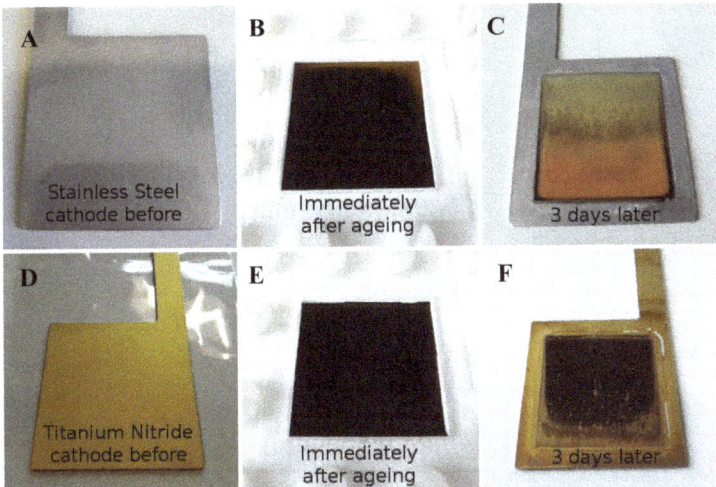

Figure 7. Visual appearance before and after Ageing Protocol A of 316 stainless-steel cathodes both coated in TiN and uncoated. (**A**) is original unaged cathode, (**B**) is immediately after aging, (**C**) is after a further 3 days sitting in 0.5 M NaOH, (**D**) is original unaged TiN coated cathode, (**E**) is immediately after aging, (**F**) is after a further 3 days sitting in 0.5 M NaOH.

3.1. SEM and EDX

The cathodes from Figure 7. were analysed using SEM and EDX, since their smaller design permitted them to be mounted inside the electron microscope, with the results as presented in Figure 8 and Table 3. The SEM micrograph showed the presence of crystals, and it was apparent from the EDX spectrum obtained that they were comprised primarily of copper. This could be attributed to the layer of copper applied underneath the TiN during the commercial titanium nitride deposition.

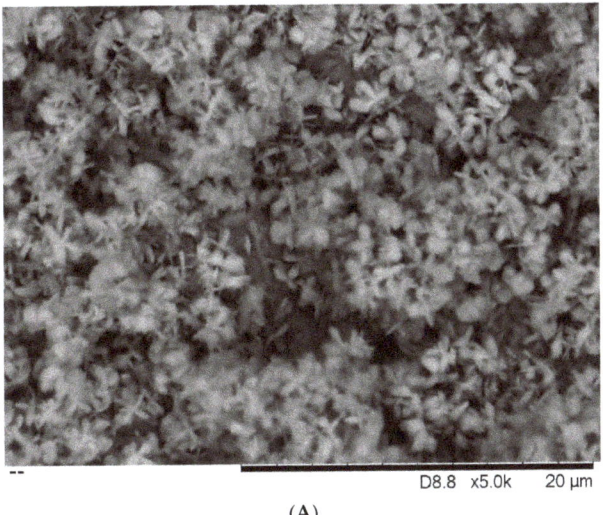

(**A**)

Figure 8. *Cont.*

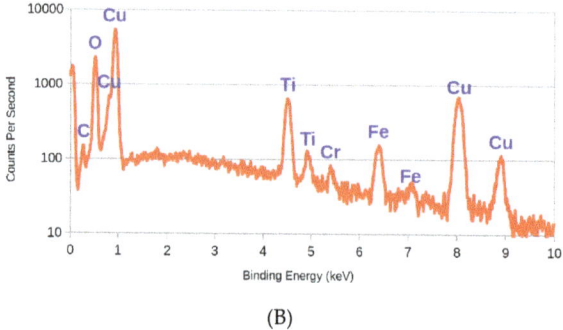

(B)

Figure 8. Electron micrograph at 5000x magnification (**A**) and EDX spectrum of the crystalline deposit (**B**) for the titanium nitride cathode after Ageing Protocol B.

Table 3. Quantitative analysis of the above EDX spectrum for the crystalline deposit.

Element	Line Type	Weight %	Sigma	Atomic %
Cu	L series	61.00	0.58	31.84
O	K series	24.07	0.38	49.90
Ti	K series	6.86	0.15	4.75
Fe	K series	3.30	0.16	1.96
C	K series	3.28	0.36	9.05
N	K series	0.90	0.64	2.13
Cr	K series	0.60	0.09	0.38
	Total			100.01

For the stainless-steel cathode, the appearance was as shown in Figure 9. There was a marked difference between the top half of the electrode (location 'a'), which appeared unaffected, and the bottom half (location 'b'), which was covered in small particles approximately 500 nm across. EDX analysis of the particles confirmed that they were 70% copper by weight.

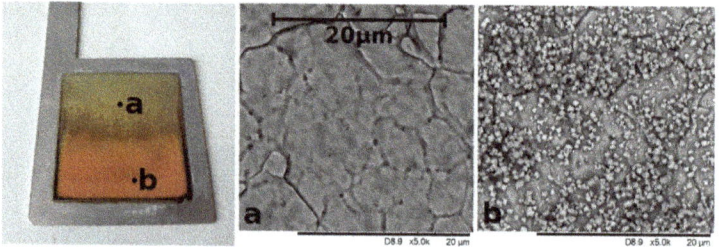

Figure 9. Electron micrographs at 5000x magnification of the stainless-steel cathode after Ageing Protocol B.

This is in agreement with the coppery colour exhibited by the electrode, but is nevertheless a surprising result, as no explicit source of copper exists in the experiment. It is therefore speculated that trace levels of copper must have been present in the electrolyte, the membrane or the stainless steel. This is supported by EDX results obtained by Kao et al. [19], and by experiments on samples of 316 grade stainless steel from two separate steel suppliers, where percentages of copper between 1 and 1.6 wt % were observed, even though according to official standards, 316 grade stainless steel does not contain copper. Regardless of the source of the contamination, this result highlights the extent to which copper can become highly concentrated on the cathode during intermittent use. There is, however, a negligible effect on the long-term performance of the electrode.

3.2. XPS (X-ray Photoelectron Spectroscopy)

The TiN coating undergoes rapid deterioration if used even briefly as an anode, as discussed in Section 3. After sweeping from 0 to 0.7 V at 10 mV/s, the coating changed to a deep orange-brown colour, and the electrical performance was drastically reduced. Examination of the original coating using XPS before and after this alteration produced results as shown in Figure 10. It is clear that there are more components present in the XPS signal of the original TiN coating.

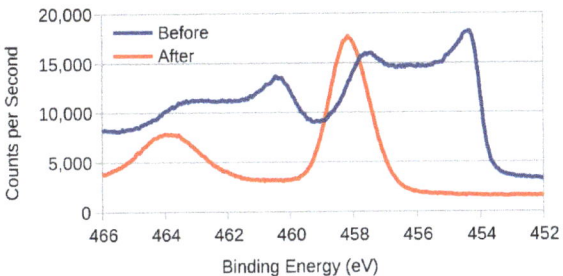

Figure 10. XPS results for the Ti 2p region before and after use as an anode.

In order to identify these components, the results were analysed using CasaXPS software, which is able to perform peak fitting. Typically, all 2p electron orbitals produce an XPS signal consisting of doublets, whereby the lower binding energy peak ($2p_{1/2}$) has double the area of the higher peak ($2p_{3/2}$), but the same full-width half-maximum (FWHM). However, it is known that the FWHM constraint is not entirely applicable to titanium, due to the Coster-Kronig effect, which causes a broadening of the $2p_{1/2}$ peak [20]. Nevertheless, it is still possible to perform outline peak deconvolution, the results of which are as presented in Figure 11.

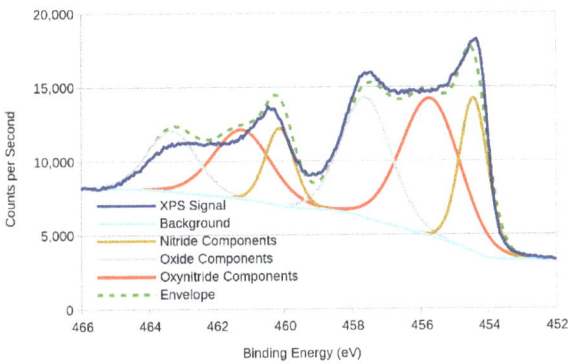

Figure 11. XPS results and component fitting for Ti 2p region of titanium nitride.

The fitting indicates the presence of three separate components within the overall XPS signal, which are accepted to correspond to the presence of titanium nitride and oxide [10,21], as well as oxynitride [17,20], as indicated. Given the positioning of the remaining component in Figure 10 (at ~458 eV), it is clear that after even brief use as an anode, the surface layers of the coating lost all traces of nitrogen and became oxidised titanium. This is understandable, since anodes are prone to oxidation in general, and titanium is prone to oxidation in particular [20]. A similar finding has been made by Wang et al. who observed that the higher the concentration of nitride in their oxynitride coating, the higher its tendency to be irreversibly oxidised under anodic conditions [22].

Since XPS is an extremely surface-sensitive analytical technique, examining exclusively the top ~10 nm, it is possible to use an ion beam to mill into the surface, and thereby obtain depth profiling information, the results of which are as shown in Figure 12. Here the numbers 1 to 9 refer to successively deeper XPS measurements and show that the nitrogen peak becomes progressively stronger with increasing depth.

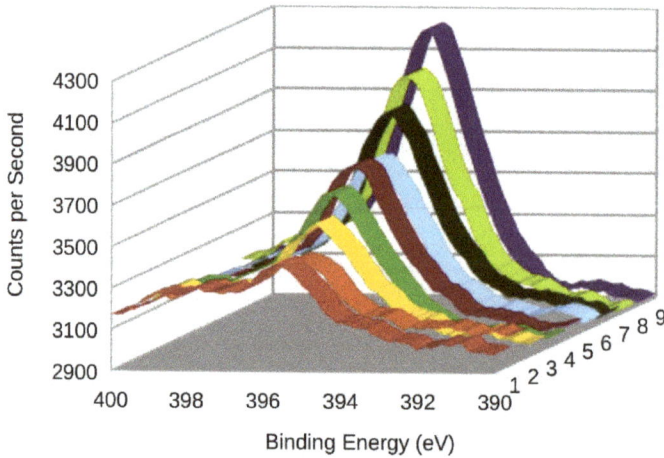

Figure 12. XPS depth profile (with smoothing) of the Ti N1s region of anodically altered titanium nitride.

The nitrogen peak reappearance indicates that the stainless steel remained protected and coated with TiN at depth, but overall the electrode incurred a loss of electrical performance caused by the oxide on the surface, which is insulating at these voltages [22]. It is therefore concluded that the phase composition of the anodically altered coating changes from TiO_x at the surface, to TiN at depth, via a mixture including titanium oxynitride, as is confirmed by the change to an orange-brown colour [23] and the changing XPS [24–26]. The practical significance of this is that incorrect wiring or fluctuations in the polarity of the electrical input would fast render the electrolysis equipment ineffective. Therefore, in practice some sort of protection would need to be provided against reverse polarity, for example via the use of a diode and fuse, or alternatively a field-effect transistor (FET). Both solutions, however, would incur some liability, either in terms of cost or reliability.

4. Conclusions

The improvement to the electrical performance and reliability of 316 grade stainless steel as a cathode for water splitting by the application of a commercially available titanium nitride coating has been investigated. Initially this appears to incur a 300 mV increase in overpotential, but over long-term intermittent experiments at 200 mA cm^{-2}, a two-electrode cell incorporating the coating was observed to outperform the uncoated material by 250 mV. At this current density, both stainless-steel and TiN cathodes experienced significant discolouration (cf. Figure 7). This appears to be partially reversible, since the deposit is observed to dissolve into the electrolyte over several days. More importantly, the coated material demonstrated a significant increase in electrical performance after such intermittent usage, improving by 400 mV, which was enough to surpass the uncoated material. Figure 6 shows that the coated material has outperformed the uncoated material in a three-electrode system, and Figure 5 shows that it has also outperformed the coated material when employed as a complete system. The SEM results in Figures 8 and 9 show that the 'black appearance' of both cathodes actually has different underlying causes.

Characterisation using SEM (scanning electron microscopy) confirmed that the migration and deposition of copper might be responsible for some of this increase. Electron micrographs of the coated material after ageing reveal a large number of sharply pointed copper crystals. It is theorised that these grew from a layer of copper that was deposited by the coating supplier before the titanium nitride coating was applied. For the uncoated material, many copper particles approximately 500 nm in diameter were observed to have been deposited. Whilst the source of this copper contamination remains unknown, their appearance is not associated with a decrease in performance.

It was also confirmed that TiN cannot be used as an anode at all, and characterisation using XPS (X-ray photoelectron spectroscopy) revealed that the coating experiences a rapid conversion to TiO_x, with the loss of all nitrogen from the surface layers. Ion beam milling revealed that the transition from TiO_x at the surface to TiN at depth is gradual, and therefore will necessarily encompass intermediary compositions of titanium oxynitride. It is possible that a ternary compound of TiAlN or CrAlN might demonstrate greater resistance to electro-oxidation, as confirmed for thermal oxidation by Chim et al. [27]. Nevertheless, this does little to detract from the applicability of TiN as a cathode for electrolytic water-splitting under intermittent room-temperature alkaline conditions.

Author Contributions: Conceptualization, C.W.D.; methodology, C.W.D.; software, W.J.F.G.; validation, W.J.F.G.; formal analysis, W.J.F.G.; investigation, W.J.F.G. and D.R.J.; resources, C.W.D.; data curation, C.W.D.; writing—original draft preparation, W.J.F.G.; writing—review and editing, W.J.F.G.; visualization, W.J.F.G.; supervision, W.J.F.G.; project administration, C.W.D.; funding acquisition, C.W.D.

Funding: PhD student William Gannon, funded by a University of Swansea Zienkiewicz Scholarship.

Acknowledgments: We would like to acknowledge the assistance provided by Swansea University College of Engineering AIM Facility, which was funded in part by the EPSRC (EP/M028267/1), the European Regional Development Fund through the Welsh Government (80708) and the Sêr Solar project via the Welsh Government. WJG thanks the Zienkiewicz Centre for his PhD funding.

Conflicts of Interest: The authors declare no conflict of interest.

References

1. Mazloomi, K.; Gomes, C. Hydrogen as an energy carrier: Prospects and challenges. *Renew. Sustain. Energy Rev.* **2012**, *16*, 3024–3033. [CrossRef]
2. Sharma, S.; Ghoshal, S.K. Hydrogen the future transportation fuel: From production to applications. *Renew. Sustain. Energy Rev.* **2015**, *43*, 1151–1158. [CrossRef]
3. Maghami, M.R.; Asl, S.N.; Rezadad, M.E.; Ale Ebrahim, N.; Gomes, C. Qualitative and quantitative analysis of solar hydrogen generation literature from 2001 to 2014. *Scientometrics* **2015**, *105*, 759–771. [CrossRef] [PubMed]
4. Gutiérrez-Martín, F.; Confente, D.; Guerra, I. Management of variable electricity loads in wind - Hydrogen systems: The case of a Spanish wind farm. *Int. J. Hydrogen Energy* **2010**, *35*, 7329–7336. [CrossRef]
5. Phillips, R.; Dunnill, C.W. Zero gap alkaline electrolysis cell design for renewable energy storage as hydrogen gas. *RSC Adv.* **2016**, *6*, 100643–100651. [CrossRef]
6. Phillips, R.; Edwards, A.; Rome, B.; Jones, D.R.; Dunnill, C.W. Minimising the ohmic resistance of an alkaline electrolysis cell through effective cell design. *Int. J. Hydrogen Energy* **2017**, *42*, 23986–23994. [CrossRef]
7. Olivares-ramírez, J.M.; Campos-cornelio, M.L.; Godínez, J.U. Studies on the hydrogen evolution reaction on different stainless steels. *Int. J. Hydrogen Energy* **2007**, *32*, 3170–3173. [CrossRef]
8. Carta, R.; Dernini, S.; Polcaro, A.M.; Ricci, P.F.; Tola, G. The influence of sulphide environment on hydrogen evolution at a stainless steel cathode in alkaline solution. *J. Electroanal. Chem.* **1988**, *257*, 257–268. [CrossRef]
9. Schäfer, H.; Beladi-Mousavi, S.M.; Walder, L.; Wollschläger, J.; Kuschel, O.; Ichilmann, S.; Sadaf, S.; Steinhart, M.; Küpper, K.; Schneider, L. Surface Oxidation of Stainless Steel: Oxygen Evolution Electrocatalysts with High Catalytic Activity. *ACS Catal.* **2015**, *5*, 2671–2680. [CrossRef]
10. Devia, D.M.; Restrepo-Parra, E.; Arango, P.J. Comparative study of titanium carbide and nitride coatings grown by cathodic vacuum arc technique. *Appl. Surf. Sci.* **2011**, *258*, 1164–1174. [CrossRef]
11. Cho, E.A.; Jeon, U.-S.; Hong, S.-A.; Oh, I.-H.; Kang, S.-G. Performance of a 1kW-class PEMFC stack using TiN-coated 316 stainless steel bipolar plates. *J. Power Sources* **2005**, *142*, 177–183. [CrossRef]

12. WallworkHT Commercial Coatings (PVD). Available online: https://www.wallworkht.co.uk/content/commercial_coatings/ (accessed on 28 January 2019).
13. Pletcher, D.; Li, X. Prospects for alkaline zero gap water electrolysers for hydrogen production. *Int. J. Hydrogen Energy* **2011**, *36*, 15089–15104. [CrossRef]
14. Passas, G.; Dunnill, C.W. Water Splitting Test Cell for Renewable Energy Storage as Hydrogen Gas. *Fundam. Renew. Energy Appl.* **2015**, *5*, 3–8.
15. Stevens, M.B.; Enman, L.J.; Batchellor, A.S.; Cosby, M.R.; Vise, A.E.; Trang, C.D.M.; Boettcher, S.W. Measurement Techniques for the Study of Thin Film Heterogeneous Water Oxidation Electrocatalysts. *Chem. Mater.* **2016**, *29*, 120–140. [CrossRef]
16. Milosev, I.; Strehblow, H.-H.; Navinsek, B. Comparison of TiN, ZrN, and CrN coatings under oxidation.pdf. *Thin Solid Films* **1997**, *303*, 246–254. [CrossRef]
17. Gebauer, C.; Fischer, P.; Wassner, M.; Diemant, T.; Jusys, Z.; Hüsing, N.; Behm, R.J. Performance of titanium oxynitrides in the electrocatalytic oxygen evolution reaction. *Nano Energy* **2016**, *29*, 136–148. [CrossRef]
18. Shinagawa, T.; Garcia-esparza, A.T.; Takanabe, K. Insight on Tafel slopes from a microkinetic analysis of aqueous electrocatalysis for energy conversion. *Sci. Rep.* **2015**, *5*, 13801. [CrossRef] [PubMed]
19. Kao, C.T.; Ding, S.J.; Chen, Y.C.; Huang, T.H. The anticorrosion ability of titanium nitride (TiN) plating on an orthodontic metal bracket and its biocompatibility. *J. Biomed. Mater. Res.* **2002**, *63*, 786–792. [CrossRef] [PubMed]
20. XPS Interpretation of Titanium. Available online: https://xpssimplified.com/elements/titanium.php (accessed on 7 June 2018).
21. Shimada, S.; Hasegawa, M. Preparation of Titanium Nitride Films from Amide Precursors Synthesized by Electrolysis. *Society* **2003**, *79*, 177–179.
22. Wang, W.; Savadogo, O.; Ma, Z.F. Preparation of new titanium oxy nitride based electro catalysts using an anhydrous sol-gel method for water electrolysis in acid medium. *Int. J. Hydrogen Energy* **2012**, *37*, 7405–7417. [CrossRef]
23. Chappé, J.M.; Martin, N.; Lintymer, J.; Sthal, F.; Terwagne, G.; Takadoum, J. Titanium oxynitride thin films sputter deposited by the reactive gas pulsing process. *Appl. Surf. Sci.* **2007**, *253*, 5312–5316. [CrossRef]
24. Dunnill, C.W.; Aiken, Z.A.; Pratten, J.; Wilson, M.; Parkin, I.P. Sulfur-and nitrogen-doped titania biomaterials via APCVD. *Chem. Vap. Depos.* **2010**, *16*, 50–54. [CrossRef]
25. Dunnill, C.W.; Ansari, Z.; Kafizas, A.; Perni, S.; Morgan, D.J.; Wilson, M.; Parkin, I.P. Visible light photocatalysts—N-doped TiO$_2$ by sol-gel, enhanced with surface bound silver nanoparticle islands. *J. Mater. Chem. A* **2011**, *21*, 11854–11861. [CrossRef]
26. Dunnill, C.W.; Parkin, I.P. N-Doped Titania Thin Films Prepared by Atmospheric Pressure CVD using t-Butylamine as the Nitrogen Source: Enhanced Photocatalytic Activity under Visible Light. *Chem. Vap. Depos.* **2009**, *15*, 171–174. [CrossRef]
27. Chim, Y.C.; Ding, X.Z.; Zeng, X.T.; Zhang, S. Oxidation resistance of TiN, CrN, TiAlN and CrAlN coatings deposited by lateral rotating cathode arc. *Thin Solid Films* **2009**, *517*, 4845–4849. [CrossRef]

© 2019 by the authors. Licensee MDPI, Basel, Switzerland. This article is an open access article distributed under the terms and conditions of the Creative Commons Attribution (CC BY) license (http://creativecommons.org/licenses/by/4.0/).

Article

Multi-Tubular Reactor for Hydrogen Production: CFD Thermal Design and Experimental Testing [†]

Elvira Tapia [1,‡], Aurelio González-Pardo [2], Alfredo Iranzo [1,*], Manuel Romero [3], José González-Aguilar [3], Alfonso Vidal [2], Mariana Martín-Betancourt [4] and Felipe Rosa [1]

1. Thermal Engineering Group, Energy Engineering Department, School of Engineering, University of Seville, Camino de los Descubrimientos s/n, 41092 Sevilla, Spain; etapia@us.es (E.T.); rosaif@us.es (F.R.)
2. CIEMAT-PSA, Carretera de Senés, S/N Tabernas, 04200 Almería, Spain; aurelioj@psa.es (A.G.-P.); alfonso.vidal@ciemat.es (A.V.)
3. IMDEA Energy Institute, Avda. Ramón de la Sagra, 3, 28935 Móstoles, Spain; manuel.romero@imdea.org (M.R.); jose.gonzalez@imdea.org (J.G.-A.)
4. ABENGOA Innovación–División de hidrógeno. C/Energía Solar,1 41014 Sevilla, Spain; mariana.martin@abengoa.com
* Correspondence: airanzo@us.es; Tel.: +34-954-487471
† This paper is an extended version of the conference paper published in SolarPACES 2016 International Conference, Santiago de Chile, Chile, 26–29 September, 2017.
‡ Present address: ABENGOA Innovación–División de hidrógeno. C/Energía Solar,1 41014 Sevilla, Spain.

Received: 3 December 2018; Accepted: 27 December 2018; Published: 11 January 2019

Abstract: This study presents the Computational Fluid Dynamics (CFD) thermal design and experimental tests results for a multi-tubular solar reactor for hydrogen production based on the ferrite thermochemical cycle in a pilot plant in the Plataforma Solar de Almería (PSA). The methodology followed for the solar reactor design is described, as well as the experimental tests carried out during the testing campaign and characterization of the reactor. The CFD model developed for the thermal design of the solar reactor has been validated against the experimental measurements, with a temperature error ranging from 1% to around 10% depending on the location within the reactor. The thermal balance in the reactor (cavity and tubes) has been also solved by the CFD model, showing a 7.9% thermal efficiency of the reactor. CFD results also show the percentage of reacting media inside the tubes which achieve the required temperature for the endothermic reaction process, with 90% of the ferrite pellets inside the tubes above the required temperature of 900 °C. The multi-tubular solar reactor designed with aid of CFD modelling and simulations has been built and operated successfully.

Keywords: solar reactor; hydrogen production; solar receiver; thermal energy; computational fluid dynamics; CFD; model

1. Introduction

The coupling of concentrated solar thermal power to industrial processes (hybridization) is a technology under development with a very high potential to reduce greenhouse gas emissions [1], but concentrated solar power can also be used for the production of fuels. The different technologies involving high temperature concentrated solar power aimed at the conversion of solar energy to chemical fuels are currently being thoroughly investigated. The main activities and efforts are focused on the endothermic reactions and processes, and on identifying and developing improved receivers and reactors for carrying out such thermochemical processes [2]. Solar reactors represent a promising technology for a future sustainable energy system. As an example, energy carriers such as hydrogen can be produced in solar reactors and stored, for its later use during the period when solar energy is not available. Also the production of liquid fuels (after a Fischer-Tropsch stage) is being investigated. Different solar reactors have been demonstrated for several chemical processes and scales.

The solar reactor is the component where solar concentrated energy is received and transformed into thermal energy. As such process is involving high radiation fluxes and high temperatures, an appropriate design of the receiver requires to reduce radiation, convection and conduction losses, as well as electrical consumption. The design must also promote the heat transfer towards the absorbent media where the active zone of the reactor is located, usually meant to host an endothermic chemical reaction. A smooth temperature profile is also desired in order to reduce thermal stresses and enhance the mechanical durability of the reactor components. Depending on the heat integration typology solar reactors are classified into indirectly or directly irradiated (Figure 1).

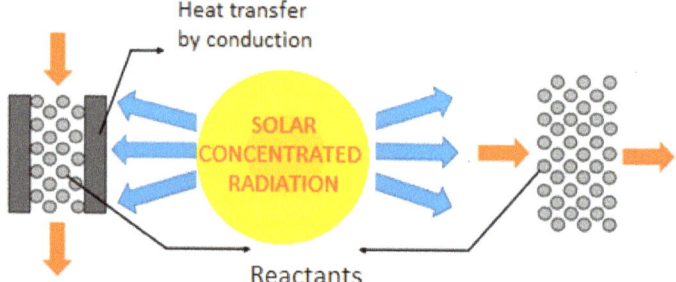

Figure 1. Conceptual sketch depicting indirect solar reactors (**left**) and direct solar reactors (**right**).

There have been several research and demonstration activities of solar reactors, covering different chemical processes and different scales. However, most processes and reactors identified in the literature are focused on methane reforming, at a laboratory scale (<1 kW) [3], pilot scale (1–100 kW) [3,4] and also commercial scale (>100 kW) [5,6]. There are other possible chemical processes such as gasification [7–10] and water splitting by means of thermochemical cycles, also at a laboratory scale [11–13], pilot scale [14,15] or commercial scale [16].

Given the significant importance of the coupling between radiation, heat transfer, fluid mechanics and chemical reactions in solar reactors, Computational Fluid Dynamics (CFD) tools are increasingly being used for the thermal design and optimization of the operating conditions. Since the early work by Meier [17] in 1996 (focused on hydrogen production with ferrites thermochemical cycles), the number of studies based on CFD simulations have significantly increased over the last years. CFD can be used to guide the design process by predicting results of temperature fields, heat flux distributions, and other variables of interest, and can be used for the main reactor types such as particle reactors, volumetric reactors and tubular reactors [18–20].

One of the most demanding issues in terms of computational times is radiation modelling. A set of different radiation models can be found in the literature for solar reactor modeling, being one of the most considered the Discrete Ordinate model (DO) [21–25]. In such studies, the boundary condition for the radiation flux is typically defined as a uniform or Gaussian distribution with a unidirectional vector [17,21–23,26–28]. However, if the radiation flux must consider the different directions of rays (which is the real case when the radiation flux is coming from a parabolic dish or a heliostat field), it is more advisable to use the Monte Carlo model (MC), as it is a more practical approach for implementing the ray directions [27–32]. When the volumetric absorption of thermal radiation can be neglected the surface-to-surface (S2S) radiation model is also used [33,34].

For the design of tubular reactors, some studies with CFD tools can be also found in the literature. In such reactor models the air movement in the cavity volume is typically considered as a laminar flow, in particular for the studies using a quartz window at the aperture. The modelling results of the studies are generally validated with experimental data, where in general a good agreement is reported (with a maximum error in temperature lower than 10%). Therefore, CFD techniques can be considered as a useful and accurate tool to be considered for solar reactor design and for the optimization of the operating conditions.

In this work, a 100 kW$_{th}$ multi-tubular cavity reactor for hydrogen production integrated in a solar tower was designed, built and tested in the framework of the SolH2 project (Hydrogen production from high temperature thermal solar energy, referenced in acknowledgements section), with the main goal to demonstrate the technological feasibility of solar thermochemical water splitting cycles as one of the most promising carbon-free options to produce hydrogen from renewable sources. This paper is an extended version of the conference paper published in SolarPACES 2016 International Conference [35].

The design of the reactor was developed to fulfill the hydrogen requirements, the temperature levels required for the process, and the corresponding inlet and outlet gas temperatures and efficiency. A CFD model and simulation analysis of the solar reactor was carried out as part of the design process in order to assess and optimize the temperature distribution and the absorbed radiation flux among the reactor tubes, as it is explained in the next sections. The experimental testing carried out at the Plataforma Solar de Almería (PSA) in Spain is also presented. The experimental measurements for the 100 kW$_{th}$ multi-tubular reactor have been used to validate the CFD model.

2. Materials and Methods

The reactor was built and installed in the SSPS-CRS (Small Solar Power Systems-Central Receiver Solar power plant) facility of the Plataforma Solar de Almería, located in Tabernas desert, Spain. The Plataforma Solar de Almeria is the largest European experimental facility on concentrating solar energy. The SSPS-CRS plant consists of an autonomous heliostats field and a 43 m height tower. The facility collects direct solar radiation by means of a field consisting of 91 heliostats, each of them with a surface of 39.3 m^2. The heliostats are distributed in 16 rows in a 150 × 70 m field. The tower is 43 m high and has two metallic testing platforms, at heights of 26 and 32 m respectively. The maximum thermal power delivered by the field onto the receiver aperture is 2.0 MW. This plant has been used in the past to perform testing of a wide variety of solar receivers and applications, in the range of 200–350 kW thermal power. The test facility has been transformed into a suitable test rig to host research initiatives in solar hydrogen production, such as HYDROSOL–Plant (Solar Hydrogen via Water Splitting in Advanced Monolithic Reactors for Future Solar Power plants), SYNPET (Solar gasification of petcoke), among others.

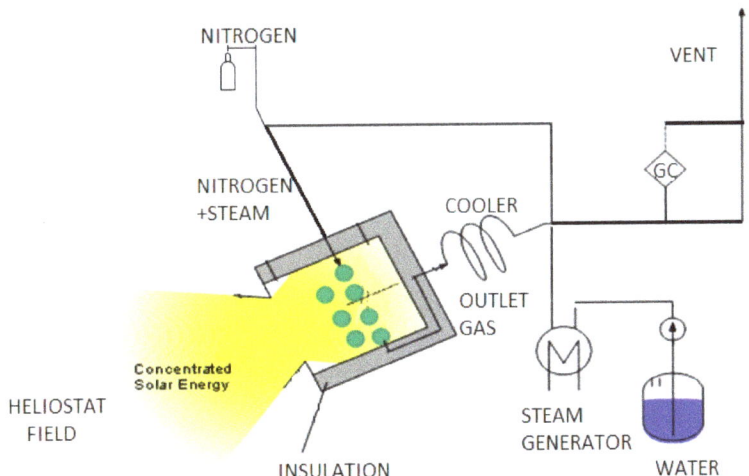

Figure 2. Sketch of the solar reactor plant.

The use of a cavity geometry for the solar reactor intends to approach a blackbody absorber in order to capture solar energy and reduce thermal emission losses (convective and radiative losses). The basic sketch of the reactor and plant is depicted in Figure 2.

Solar radiation is concentrated from the heliostat field into the receiver aperture. It spreads inside the cavity and is absorbed by the alumina tubes and also by the internal walls, where heat is also transferred to the tubes by conduction. A carrier gas enters the reactor by the upper part manifold, flowing downwards along the tubes which are filled up with ferrite pellets. The outlet gas circulates towards a heat exchanger used to preheat inlet gases, and is then cooled down in a second heat exchanger in order to precondition the gas temperature for the chromatography system. The outlet gases are not stored, and a small sample is directed to the chromatograph to analyze the composition.

2.1. Solar Reactor

According to the hydrogen production specifications and the initial geometry conditions, a first design for the solar reactor was proposed. The initial design consisted of a 2 m radius semi-cylindrical cavity with a square opening of 30 × 30 cm^2. Inside the cavity there were 80 tubes positioned in a staggered arrangement in 2 rows (Figure 3, left). That design was finally dismissed due to the following reasons: first, the empty space in the middle of the rows in front of the opening, and secondly the too small aperture for the incident radiation delivered by the heliostat field, which resulted in an increase of the temperature difference within tubes and the spillage at the aperture.

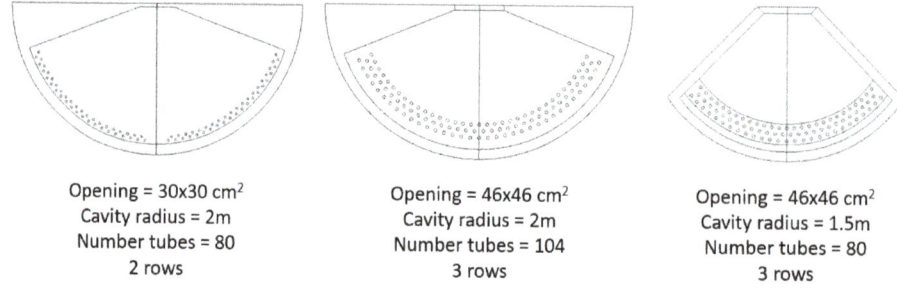

Figure 3. (**left**) Initial reactor design; (**middle**) Second reactor design; (**Right**) Final reactor design.

The next reactor version was designed in order to avoid the issues found in the original design. The dimensions of the opening were increased up to 46 × 46 cm^2 and the tubes distribution was modified. Tubes were placed in 3 rows in a staggered arrangement, increasing the number of tubes up to 104. The tubes located at both extremes, receiving less direct radiation, were intended to be used as a preheating heat exchanger (Figure 3, middle). Despite the improvements, this design was also dismissed, as the larger cavity radius involved a significantly slower transient thermal heating, and the efficiency of the preheating exchangers at both extremes was low.

The final design consists of a semi-cylindrical shape, with a 1.5 m internal cavity radius and 80 tubes positioned in three rows in staggered arrangement. The design included tubes in front of the opening (Figure 3, right).

All design steps were modelled and simulated with CFD tools (ANSYS-CFX software, Version 14.0, ANSYS Inc., Canonsburg, PA, USA) using a Monte Carlo surface-to-surface radiation model, in order to analyze the incident radiation flux on the walls and on the tubes, and the resulting temperature distributions and heat transfer to the process side of the reactor (the interior of the tubes). The final design was built as result of the SolH2 project (see Figure 4) and a thermal characterization of the reactor was carried out. The tubes have a length of 1.2 m and a total volume of 566 cm^3. Tubes are filled with small pellets of mixed ferrites, cylindrically shaped in order to increase the reaction surface and enhance the gas flow through them.

Figure 4. (**left**) Frontal view of the reactor installed in CRS (Central Receiver Solar) tower plant; (**right**) View of the reactor being irradiated by CRS field.

2.2. Flux Measurement System

The radiation flux measuring system is used to characterize the power arriving to the aperture of the cavity. It is based on capturing the irradiance distribution on a moving Lambertian target with a high resolution CCD (couple-charged device) camera [36]. The bar intercepts the concentrated beam in the measuring plane, which is located as close to the receiver as possible. The distribution of relative intensity is recorded as gray-scale map, and represents the shape of the flux distribution at the receiver´s aperture. A radiometer is used to calibrate the system, where the gray-scale values of the pixels of the image are correlated with the corresponding irradiance value measured by the radiometer.

Temperatures at different locations within the reactor are also measured. Four thermocouples are included inside the reactor tubes in order to characterize the thermal behavior of the tubes, and also to compare the experimental tests with the CFD results. The thermocouples are installed in the middle of the tube (at 0.6 m), for tube 14 and tube 21 in the first row, and tube 3 and 12 in the second one, as depicted in Figure 5.

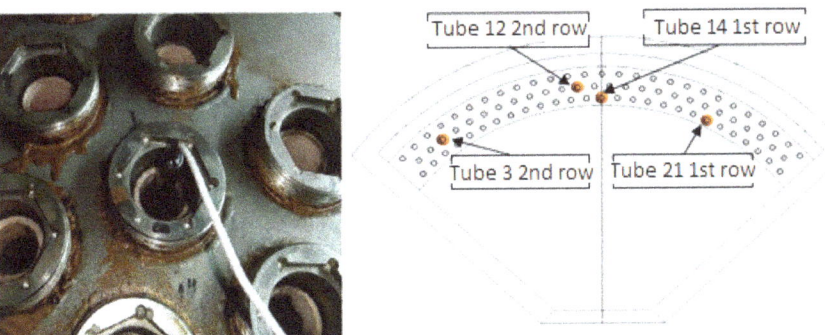

Figure 5. (**left**) Detailed view of the tubes filled with ferrites and with a thermocouple installed; (**right**) Distribution of the four thermocouples arranged inside the tubes.

2.3. Thermal Tests

The preliminary experiences with alumina tubes showed that they were very sensitive to temperature gradients. Some tubes were broken in previous tests due to thermal stresses caused by temperature gradients along the tube length. Such evidences, together with the reactor design requirement for a homogenous temperature distribution, made necessary an appropriate strategy for the heliostats field. The final strategy used was the following: first, only two groups of heliostats were focused on the aperture, and then the number of groups was increased until the required reactor temperature was achieved. With 20 heliostats focused, a total power of 44.1 kW can be delivered inside the cavity and a temperature of 750–800 °C is reached. Finally, with 36 heliostats and an incident power about 80 kW, 1200 °C were reached [37]. As the thermochemical cycle is based on two steps (activation and hydrolysis) this solar reactor design concept implies a discontinuous hydrogen production over the operation time. Therefore, a sequential mode of operation was implemented in order to couple the operation of the plant and heliostat field with the mixed ferrites thermochemical cycle, which consists of two reaction stages.

The study presented here was carried out the vernal equinox, which is a representative day of the year. The test was performed with nitrogen and without water vapor (thus without hydrogen production), with a total mass flow in the tubes of 40 kg/h and gas inlet temperature of 70 °C. The exact procedure to focus the groups of heliostats is described below (and depicted in Figure 6), and it was designed to enable the solar radiation to impinge on all the alumina tubes: first, groups 1 and 2; then group 3 and then continued with groups 4 and 5. The power ramp was done progressively without waiting for the stabilization of the tubes temperature. Then, groups 10 and 11 were included and finally groups 8 and 9 (Figure 6). All groups contain both nearby and distant heliostats in order to favor the homogeneity of the radiation flux [37]. Once the final configuration of the heliostats was active, the incident radiation flux and direction of rays at the opening were measured as described in Section 2.2. Such measurements were used as radiation boundary conditions for the definition of the CFD model and simulations.

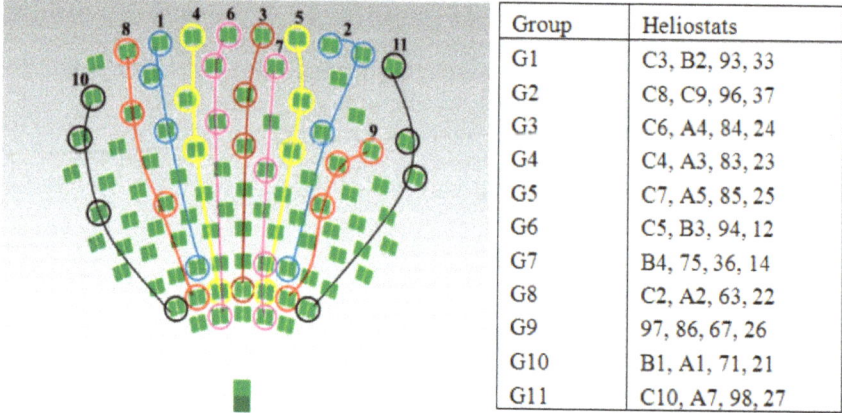

Figure 6. Groups of heliostats in which the field has been divided.

Group	Heliostats
G1	C3, B2, 93, 33
G2	C8, C9, 96, 37
G3	C6, A4, 84, 24
G4	C4, A3, 83, 23
G5	C7, A5, 85, 25
G6	C5, B3, 94, 12
G7	B4, 75, 36, 14
G8	C2, A2, 63, 22
G9	97, 86, 67, 26
G10	B1, A1, 71, 21
G11	C10, A7, 98, 27

2.4. Computational Fluid Dynamics (CFD) Model

The commercial software ANSYS-CFX 14.0 [38] has been used for the modelling and resolution of the fluid flow and heat transfer in the solar multi-tubular reactor. The geometry of the reactor and the definition of the operating conditions (radiation flux at opening and mass flow and temperature for the inert gas) are required for the model. The modeling methodology described in Tapia et al. [39] was used for the development of the CFD model. The methodology was derived for tubular solar reactors. Fluid-dynamics model for the air within the cavity and carried gas flowing along the tubes is considered, as well as thermal models for al processes involved (conduction, convection and radiation). This methodology proposes the Shear Stress Transport (SST) turbulence model for the air within the cavity when no quartz window is used, due to its better accuracy for the thermal boundary layer resolution. The Monte Carlo model is proposed for the radiation modelling, in order to enable the definition of a matrix of rays at the opening (accounting for the real directions of the radiation distribution at the opening). The S2S (surface-to-surface) model is defined as the media (air) is a nonparticipating media (no volumetric absorption of thermal radiation). There are two main modelling domains, as described below:

- Cavity: the design of the cavity is focused on achieving both a high optical efficiency and a uniform temperature distribution. The minimum temperature must be greater than the required temperature for the hydrogen production process. A mesh and a radiation factor (number of histories in Monte Carlo model) independence analysis are included in the model development. The radiation factor independence analysis is required in order to ensure that the number of histories used in the Monte Carlo model is enough to correctly calculate the radiation and temperature field (i.e., a higher number of histories do not influence the results). The cavity model includes the thermal insulation to avoid losses.
- Reactor (tubes): the reactor design starts once the cavity design is closed. The main goals are a complete thermal balance of the reactor, identifying hot/cold spots, and determining the reaction volume which achieves the required conditions. The tubes are modelled considering its thickness, and the ferrite pellets in the reactor interior are modelled as a porous media.

For a properly defined CFD analysis, it is necessary to previously carry out a mesh analysis in order to ensure that results are independent of the mesh used in the simulations. A mesh independence analysis was carried out at the beginning of the simulation process in order to identify the optimized computational mesh. The mesh is refined near wall and high gradient regions (for both temperature and velocity values). After mesh and radiation factor analysis were done, it was concluded that the final mesh featured 1.9 million nodes and 9.0 million elements. The final mesh is depicted in Figure 7. A hybrid mesh has been used, with tetrahedral elements in the cavity volume and hexahedral elements in the tube interior and wall thickness. A set of prism layers was placed at the fluid side of the reactor and tubes walls in order to ensure the appropriate resolution of the viscous and thermal boundary layers. The heat conduction through the tube wall thickness was resolved with three hexahedral elements.

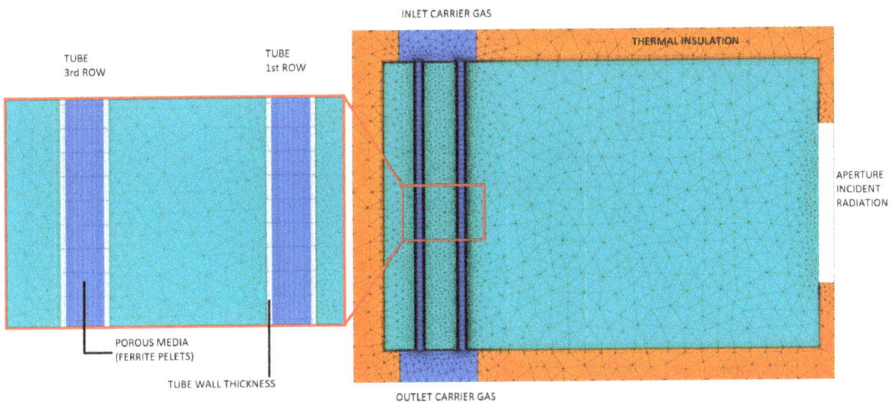

Figure 7. CFD mesh used for the solar reactor simulation (tubes detail shown on the left).

The minimum radiation factor (number of histories in the Monte Carlo model) determined to be required for this reactor is 65×10^6, which is clearly higher than the default value of 1×10^4. The value of 65×10^6 histories was determined from the radiation factor independence analysis (depicted in Figure 8), where the standard deviation of the irradiation heat flux (%SD) over the reactor surfaces is calculated as a function of the number of histories. The parameter %SD reveals the degree of accuracy of the radiation calculation in the Monte Carlo model [38], where a value less than 30% is required for an appropriate accuracy [38]. The Monte Carlo radiation solver computes the standard deviation error based on Poisson statistics. The user-specified number of histories is divided into several groups. Histories are selected from each group and their physical interactions (emission, absorption, reflection) are tracked through the domain. At the end of the calculation, each group provides values for the quantities of interest, such as irradiation heat flux or absorbed radiation. The mean value and standard deviation of each quantity of interest are computed from the groups. The normalized standard deviation (parameter %SD) is computed by dividing the standard deviation by the mean value. Figure 8 shows that 65×10^6 histories were required for reducing the %SD parameter below 30%, and thus this value was used for the simulations. Obviously, a higher number of histories requires a higher computational time for the resolution of the model. This is also presented in Figure 8.

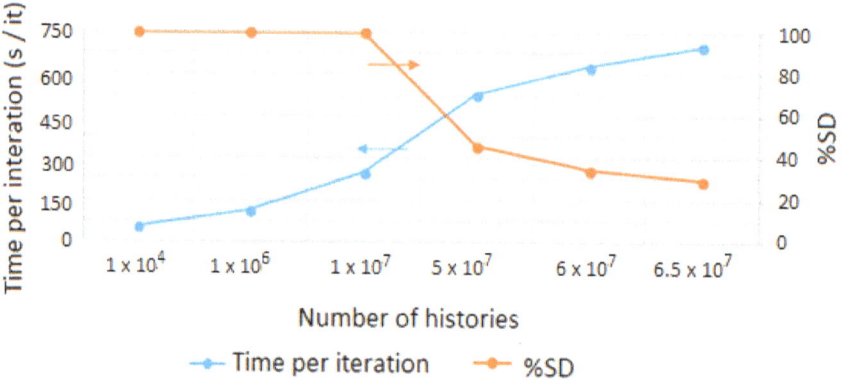

Figure 8. Results of the radiation factor (number of histories in Monte Carlo model) independence analysis.

The concentrated solar radiation enters the cavity receiver through the opening. The incident radiation flux and direction vectors were provided by CIEMAT/PSA according to the experimental tests and flux measurements as described in Sections 2.2 and 2.3. The direction of the rays, as implemented in the CFD model, is presented in Figure 9. The total solar power entering the cavity is 80 kW.

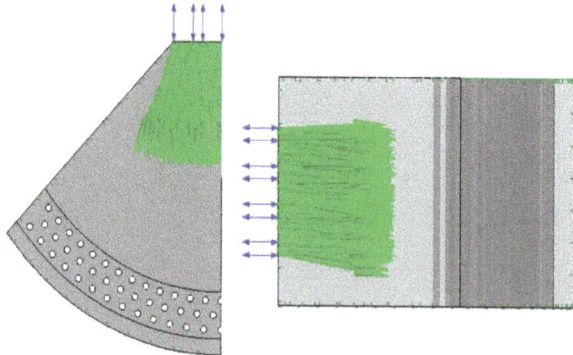

Figure 9. Direction of radiation reflected from CRS field (36 heliostats). Plant view (**left**); Side view (**right**).

The carrier gas (nitrogen) is fed to the inlet collector located on the top of the reactor, and flows downwards along the tubes finally reaching the outlet collector. The fluid flow inside the cavity and tubes, air and nitrogen respectively, is assumed to be turbulent and laminar respectively. Ideal gas approximation is use in the model, and air is defined as transparent to radiation, so that surface-to-surface radiation model is used without volumetric absorption. The reactor walls are considered as opaque surfaces, diffusely emitting and reflecting. The Monte Carlo model with surface to surface radiation model is used. The heat transfer by conduction within the tubes and porous media (ferrite pellets) is included in the same simulation where the radiation behavior is analyzed (and also heat transfer by convection between carrier gas and porous media within the tubes), therefore avoiding the need for decoupling the models. Thus, all relevant thermal processes are included in a single model and simulation. The following physical properties were considered: specific heat capacity of the nitrogen gas at constant pressure is 1041 J/kg·K and dynamic viscosity, density and thermal conductivity were defined as functions depending on temperature. The pressure drop in the porous media (ferrite pellets within the tubes) was modelled by Darcy's law with linear and quadratic coefficients as introduced in the previous model by the authors [39]. Pressure drop was 50 kPa. The thermal conductivity of silicon carbide and alumina are also defined as functions depending on temperature. The surface emissivity of the diffusely-reflecting cavity inner wall (silicon carbide) and the tubes (alumina) are 0.3 and 0.8 respectively. The nitrogen gas boundary conditions (flow rate and temperature) were defined in order to replicate the experimental test (Section 2.3). The effect of the cavity insulation at the external walls was included in the model by defining an external heat transfer coefficient of 0.128 W/m²·K (which was calculated to be equivalent to the insulation material heat conduction resistance and external heat convection resistance). All simulations were carried out in steady-state mode, thus considering nominal operating conditions and no transient effects.

The summary of physical models and boundary conditions used in the simulation is presented in Table 1.

Table 1. Physical models and boundary conditions used in the CFD (Computational Fluid Dynamics) model.

	Model/Boundary Condition Value
Radiation model (cavity air)	Monte Carlo Surface-to-Surface
Turbulence model (cavity air)	Shear Stress Transport (SST)
Turbulence model (N_2 tubes porous media)	Laminar
Ferrite porosity in tubes	0.4
Cavity window (air)	Opening Boundary Condition [38]
Cavity window (radiation)	80 kW with ray matrix profile (direction and intensity) from experimental data
Tubes emissivity	0.4 (alumina)
Receiver walls emissivity	0.9 (silicon carbide)
Receiver external walls	0.128 W/m²·K, 25 °C
Nitrogen tubes inlet	40 kg/h; 70 °C

All simulations were carried out in a HP Z600 workstation, running on parallel on 8 processors. The simulation time for the final mesh used was in the range of 5 days per simulation.

3. Results and Discussion

The results of the thermal tests and the comparison between the experimental measurements and the CFD model results is shown in this section. The time-evolution of the reactor temperatures during the experimental test until achieving the steady state for the final configuration of the heliostat field (80 kW) is presented in Figure 10. The steady state CFD results are also presented in Figure 10 (thus, for the final time only). The black stepwise curve in Figure 10 represents the number of heliostats used, indicating the progressive use of the heliostats during the start-up of the reactor, as described in Section 2.3. Black dots represent the thermal power delivered to the cavity as a results of the increasing number of heliostats being used. Colored lines represent the experimental time-evolution of the reactor temperatures during the heating-up process until reaching steady state conditions. Finally, the set of four colored dots at the final time represent the CFD results for the tubes temperature corresponding to the experimental thermocouple locations.

It can be observed that the total heating-up ramp is taking 4 h, where during the last 1.5 h all the heliostats involved are focused.

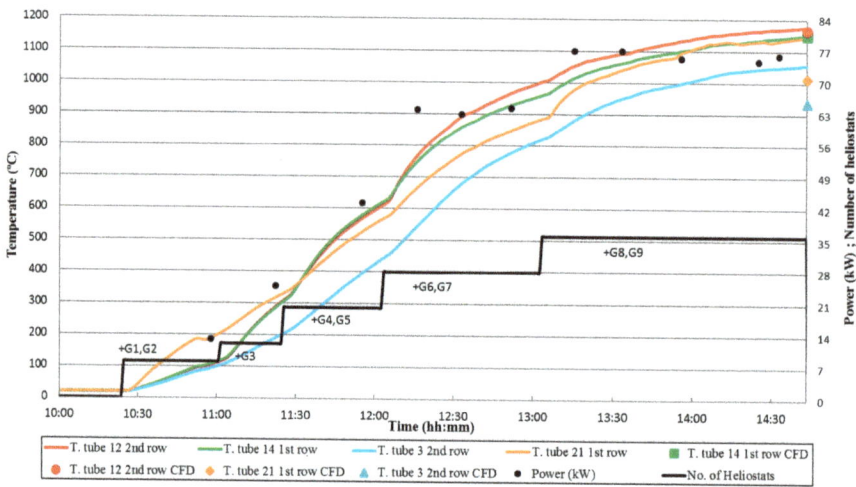

Figure 10. Evolution of temperatures for experimental test and CFD temperature results. T. tube corresponds to the thermocouple inside the tubes, in contact with ferrites. Number of heliostats focused on the receiver and the power measured.

The maximum temperature differences between experimental thermocouples and CFD results is featured by tube 21 for the first row and tube 3 in the second row. These tubes are located close to the cavity extremes so the incident radiation flux is lower than at the middle cavity tubes. A further analysis for the direction radiation flux defined in the model and also the details of the mesh at the opening (where the rays direction is defined) needs to be done.

The detailed experimental and CFD results are shown in Table 2. It is observed that the CFD temperature results are lower than the experimental temperature data for all thermocouples. This fact could be caused by larger convection losses in the CFD simulations than in the real experimental tests, as well as difference between the considered and real materials emissivity. However, the CFD model results for the central tubes (12 1st row and 14 2nd row) are presenting an excellent agreement with respect to the experimental values. The maximum temperature error in CFD simulations is around 11.5% which is in the same range as others studies [24,27,29,32]. It can be also observed that errors are very small at central tubes, and larger at the side tubes. The reason for this difference is under investigation. As the incident radiation received by central and side tubes is not the same (mainly direct incident radiation from the heliostat field for central tubes, and reflections and receiver emissions for side tubes) it is possible that either the material emissivity or the accuracy of the radiation model is not fully appropriate for representing the real radiation field within the cavity. This is however still under investigation in order to better assess the reason for this behavior.

Table 2. Comparison of experimental temperature and CFD results.

Thermocouple Location	Experimental Test (°C)	CFD Results (°C)	Error (%)	Difference (K)
Tube 12 (1st row)	1178	1167	0.1	1
Tube 14 (2nd row)	1151	1150	0.9	11
Tube 21 (1st row)	1147	1015	11.5	132
Tube 3 (2nd row)	1058	938	11.3	120

The CFD simulations also allow to evaluate the temperature distribution in tubes. Figure 11 (top) shows that temperature in tubes located at the cavity edges are lower than the rest of the tubes. Outer wall temperatures of the tubes are ranging from less than 900 °C at the top of the edge tubes to over

1200 °C at the middle of the tubes located at the centre of the cavity. The effect of the carrier gas inlet temperature is also observed, as the upper part of tubes is clearly presenting colder temperatures, and then is quickly heated up due to the heat transfer from the incident solar radiation. Figure 11 (bottom) depicts the average temperature at the outer surface of each tube. Average temperature is above 900 °C for all tubes. A Gaussian temperature distribution is clearly observed, with higher temperatures at the center due to the peak incident radiation. In general, temperatures are higher for the first row (receiving direct radiation), with temperatures of the second row nearly as high as for the first row (due to the staggered arrangement of the tubes). Third row is clearly presenting a lower temperature as the shadow effect of the first and second rows is significant and there is less direct incident radiation. Additionally, other differences between tube rows can be observed. Near the cavity extremes, the behavior of the second row is approaching the behavior of the third row. This is because at this locations direct radiation is becoming less pronounced, and secondary reflections and emission from nearby hot surfaces is governing the radiation heat transfer. This is not the case for the tubes at the center of the cavity, where first and second row receiving a direct radiation flux are presenting very similar temperatures, whereas third row is around 30 °C colder due to the shadow effect mentioned above. On average, temperature differences among the different tubes is around 120–140 °C, and thus temperature distribution can be considered relatively homogenous inside the cavity for the purpose of the endothermic process.

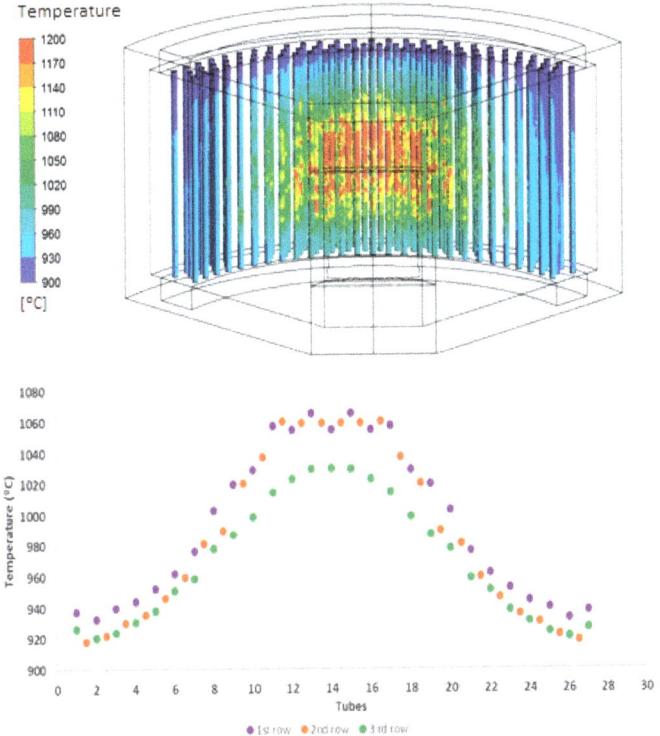

Figure 11. (**top**) Temperature distribution over the tubes in the CFD simulations; (**bottom**) Distribution of the average temperature over the tubes.

The total volume of the ferrite pellets achieving the required temperature level for the endothermic reaction has been calculated from the results of the CFD simulation. This is graphically presented in Figure 12. This is an indication of the efficiency of the reaction volume (i.e., how much volume of the reactor is actually active for the chemical reaction). In this case, simulation results show that 90.4% of the ferrite domain achieves the required process temperature (900 °C). Only the ferrite pellets at the top corners at both sides of the reactor are not achieving the required process temperature (Figure 12). The total reactor volume efficiency could be further enhanced with a better energy integration of the reactor and system, for example by exchanging energy of the outlet nitrogen gas flow with the inlet nitrogen flow, in order to preheat this stream and feed the reactor tubes with a preheated carrier gas flow. Apart from achieving higher temperatures at the upper part of the tubes, mechanical and thermal stresses would be also reduced, thus enlarging the life time of the reactor.

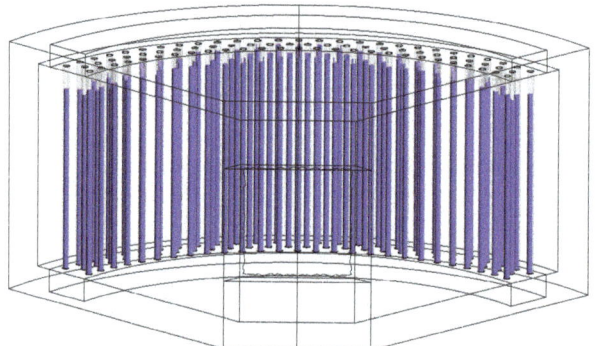

Figure 12. Volume of ferrite (inside tubes) where temperature is higher than 900 °C.

It must be considered that the carrier gas flow distribution among the 80 tubes is also relevant for the final performance of the reactor. A highly non-uniform distribution could lead to a lack of carrier gas in tubes with less flow, or to an unnecessary cooling at the tube entrance in tubes with excessive flow. The pressure drop caused by the ferrite pellets is expected to contribute to the uniformity of the gas flow among the different tubes, but in order to verify this the mass flux per tube can be analyzed from the CFD results. This is presented in Figure 13, where a slight non-uniformity is observed, with higher gas flow at the side tubes and lower gas flow in the central tubes. However, differences are less than 10% and this has been considered as acceptable. The reason for the slight non-uniformity is caused by the properties of the carrier gas (density and viscosity). The side tubes are colder than the central tubes as observed in Figure 11, thus with a higher gas density and lower viscosity. The effect of this is that pressure drop is reduced along the side tubes with respect to the central tubes, and this is causing that a slightly higher gas mass flow is processed by the side tubes.

Figure 13. Carrier gas (nitrogen) mass flux distribution over the 80 tubes of the solar reactor.

The thermal balance of the reactor is presented in Table 3. Both convection and radiation losses at the opening of the cavity have been calculated as a result of the CFD modelling. Convection losses represent 13 kW (16% of the total incoming thermal power) and radiation losses 19 kW (accounting for 24% of the total power). Finally, 6.3 kW can be transferred by the reactor towards the reaction media inside the tubes, representing a thermal efficiency of 7.9% for the solar reactor.

Table 3. Thermal balance in solar reactor.

Balance Variable	Definition	Results (CFD)
Total Incident radiation (kW)	Data	80.00
Opening radiation losses (kW)	Data	19.03
Opening convection losses (kW)	-	12.82
Heat transferred to tubes (kW)	-	6.34
Optical efficiency (%)	$\eta_{op} = \frac{IncRad - Rad\ losses}{IncRad}$	76.21
Thermal efficiency (%)	$\eta_t = \frac{Heat\ transfer\ to\ process}{IncRad}$	7.93

Thermal efficiency has been defined as the ratio between useful heat (heat transferred to the process, i.e., inner side of the reactor tubes) and incoming radiation into the reactor. Optical efficiency is defined as the ratio between radiation absorbed by the tubes and the incoming radiation. The latter mainly depends on reactor geometry and optical properties. Regarding thermal efficiency, values reported in the literature are in a very wide range. For instance, Ma et al. [40] obtained thermal efficiencies between 27% and 90% depending on working conditions such as the gas flow rate. Values reported in Table 3 show that conduction heat losses through the reactor walls represent 52.26% of the incoming radiation power, where the opening convection and radiation heat losses represent 39.81%, and the useful heat to the process 7.93%. It is therefore concluded that there is a clear room for increasing thermal efficiency by reducing conduction heat losses (i.e., improvement of reactor isolation) in order to enhance the overall mean temperature in the cavity.

4. Conclusions

A multi-tubular solar reactor for hydrogen production by thermochemical cycle has been designed with the aid of CFD modeling and simulations. The reactor has been built and thermal experimental evaluations were carried out at the Plataforma Solar de Almería (PSA). CFD results regarding the analysis of the thermal behavior are presented in this work, as well as the results of the experimental thermal characterization, for the investigation of the thermal receiver performance and its operational behavior under the operating conditions specified. The operation temperatures are ranging from 800 to 1200 °C according to the specific thermochemical cycle based on ferrites. An operation strategy of the heliostat field was previously developed by ray tracing simulations, in order to supply the required power with an optimal radiation flux distribution over the reactor tubes. The CFD model was developed following an established methodology and used for the calculation of the temperatures and radiation flux distributions at the cavity walls and tubes. The results have been validated against the data obtained during the thermal testing experimental campaign, obtaining a good agreement for tubes temperature at the center for the cavity, and around 10% temperature differences for the tubes located at the extremes of the cavity. In addition, the temperature distribution within the tubes was calculated in order to analyze its degree of uniformity and whether the required process temperature is achieved. It was shown that 90% of the reactor volume (ferrite pellets) achieve the required temperature, and further energy integration proposals are identified in order to improve this value. The scope of the analysis is to assess the thermal efficiency as well as the temperature distribution over the receiver. The CFD model provides useful information for the assessment of design parameters and to optimize the thermal performance of the solar cavity.

Author Contributions: Conceptualization, A.G.-P., A.I., M.R., J.G.-A., A.V., M.M.-B.; methodology, E.T., A.G.-P., A.I., M.R., J.G.-A., A.V.; software, E.T., A.I.; validation, E.T., A.G.-P., A.I., A.V.; formal analysis, E.T., A.G.-P., A.I.; investigation, E.T., A.G.-P., A.I.; M.R., J.G.-A.; resources, A.V., M.R., J.G.-A., M.M.-B., F.R.; data curation, E.T., A.G.-P.; writing—original draft preparation, E.T., A.G.-P., A.I.; writing—review and editing, A.I.; visualization, E.T., A.G.-P., A.I.; supervision, A.I., M.M.-B., F.R.; project administration, M.R., M.M.-B., F.R.; funding acquisition, M.R., M.M.-B., F.R.

Funding: The authors would like to thank the funding received from the Spanish Ministry of Science and Innovation (National Plan for Scientific Research, Development and Technological Innovation, 2008–2011), and the European Regional Development Fund (ERDF) for the financial support given to the SolH2 project (grant IPT-2011-1323-920000), as well as the FEDER Operational Program for Andalusia 2007–2013 for the financial support given to the RNM-6127 project.

Conflicts of Interest: The authors declare no conflict of interest.

References

1. Stuber, M.D. A Differentiable Model for Optimizing Hybridization of Industrial Process Heat Systems with Concentrating Solar Thermal Power. *Processes* **2018**, *6*, 76. [CrossRef]
2. Steinfeld, A. Solar thermochemical production of hydrogen—A review. *Sol. Energy* **2005**, *78*, 603–615. [CrossRef]
3. Dahl, J.K.; Buechler, K.J.; Weimer, A.W.; Lewandowski, A.; Bingham, C. Solar-thermal dissociation of methane in a fluidwall aerosol flow reactor. *Int. J. Hydrogen Energy* **2004**, *29*, 725–736. [CrossRef]
4. Levitan, R.; Rosin, H.; Levy, M. Chemical reactions in a solar furnacedirect heating of the reactor in a tubular receiver. *Sol. Energy* **1989**, *42*, 267–272. [CrossRef]
5. Worner, A.; Tamme, R. CO_2 reforming of methane in a solar driven volumetric receiver reactor. *Catal. Today* **1998**, *46*, 165–174. [CrossRef]
6. Epstein, M.; Spiewak, I. Solar experiments with a tubular reformer. In Proceedings of the International Symposium on Solar Thermal Concentrating Technologies, Koeln, Germany, 6–11 October 1996; pp. 1209–1229.
7. Trommer, D.; Hirsch, D.; Steinfeld, A. Kinetic investigation of the thermal decomposition of CH_4 by direct irradiation of a vortex-flow laden with carbon particles. *Int. J. Hydrogen Energy* **2004**, *29*, 627–633. [CrossRef]
8. Rodat, S.; Abanades, S.; Sans, J.-L.; Flamant, G. Hydrogen production from solar thermal dissociation of natural gas: Development of a 10 kW solar chemical reactor prototype. *Sol. Energy* **2009**, *83*, 1599–1610. [CrossRef]
9. Kruesi, M.; Jovanovic, Z.R.; Steinfeld, A. A two-zone solar-driven gasifier concept: Reactor design and experimental evaluation with bagasse particles. *Fuel* **2014**, *117*, 680–687. [CrossRef]
10. Dunn, R.; Lovegrove, K.; Burgess, G. *Ammonia Receiver Design for a 500 m^2 Dish*; Australian National University (ANU): Perpignan, France, 2010.
11. Gokon, N.; Murayama, H.; Nagasaki, A.; Kodama, T. Thermochemical two-step water splitting cycles by monoclinic ZrO_2-supported $NiFe_2O_4$ and Fe_3O_4 powders and ceramic foam devices. *Sol. Energy* **2009**, *83*, 527–537. [CrossRef]
12. Gokon, N.; Hasegawa, T.; Takahashi, S.; Kodama, T. Thermochemical two-step water-splitting for hydrogen production using Fe-YSZ particles and a ceramic foam device. *Energy* **2008**, *33*, 1407–1416. [CrossRef]
13. Abanades, S.; Flamant, G. Thermochemical hydrogen production from a two-step solar-driven water-splitting cycle based on cerium oxides. *Sol. Energy* **2006**, *80*, 1611–1623. [CrossRef]
14. Bertocchi, R.; Karni, J.; Kribus, A. Experimental evaluation of a non-isothermal high temperature solar particle receiver. *Energy* **2004**, *29*, 687–700. [CrossRef]
15. Kaneko, H. Solar hydrogen production with rotary-type solar reactor in international collaborative development between Tokyo Tech and CSIRO. In Proceedings of the 15th Solar PACES International Symposium, Berlin, Germany, 15–17 May 2009.
16. Agrafiotis, C.; Roeb, M.; Konstandopoulos, A.G.; Nalbandian, L. Solar water splitting for hydrogen production with monolithic reactors. *Sol. Energy* **2005**, *79*, 409–421. [CrossRef]
17. Meier, A.; Ganz, J.; Steinfeld, A. Modeling of a novel high temperature solar chemical reactor. *Chem. Eng. Sci.* **1996**, *51*, 3181–3196. [CrossRef]
18. Dersch, J.; Mathijssen, A.; Roeb, M.; Sattler, C. Modelling of a solar thermal reactor for hydrogen generation. In Proceedings of the 5th International Modelica Conference, Vienna, Austria, 4–5 September 2006.
19. Melchior, T.; Perkins, C.; Weimer, A.W.; Steinfeld, A. A cavity receiver containing a tubular absorber for high-temperature thermochemical processing using concentrated solar energy. *Int. J. Therm. Sci.* **2008**, *47*, 1496–1503. [CrossRef]
20. Villafan-Vidales, H.I.; Abanades, S.; Caliot, C.; Romero-Paredes, H. Heat transfer simulation in a thermochemical solar reactor based on a volumetric porous receiver. *Appl. Therm. Eng.* **2011**, *31*, 3377–3386. [CrossRef]

21. Costandy, J.; El Ghazal, N.; Mohamed, M.T.; Menon, A.; Shilapuram, V.; Ozalp, N. Effect of reactor geometry on the temperature distribution of hydrogen producing solar reactors. *Adv. Hydrogen Prod.* **2012**, *37*, 16581–16590. [CrossRef]
22. Chen, H.; Chen, Y.; Hsieh, H.-T.; Siegel, N. Computational fluid dynamics modeling of gas-particle flow within a solid-particle solar receiver. *ASME* **2007**, *129*, 160–170. [CrossRef]
23. Ozalp, N.; JayaKrishna, D. CFD analysis on the influence of helical carving in a vortex flow solar reactor. *Int. J. Hydrogen Energy* **2010**, *35*, 6248–6260. [CrossRef]
24. Abanades, S.; Flamant, G. Experimental study and modeling of a high-temperature solar chemical reactor for hydrogen production from methane cracking. *Int. J. Hydrogen Energy* **2007**, *32*, 1508–1515. [CrossRef]
25. Martinek, J.; Bingham, C.; Weimer, A.W. Computational modeling of a multiple tube solar reactor with specularly reflective cavity walls. Part 2: Steam gasification of carbon. *Chem. Eng. Sci.* **2012**, *81*, 285–297. [CrossRef]
26. Roldan, M.I.; Canadas, I.; Casas, J.L.; Zarza, E. Thermal analysis and design of a solar prototype for high-temperature processes. *Int. J. Heat Mass Transf.* **2013**, *56*, 309–318. [CrossRef]
27. Hirsch, D.; Steinfeld, A. Solar hydrogen production by thermal decomposition of natural gas using a vortex-flow reactor. *Int. J. Hydrogen Energy* **2004**, *29*, 47–55. [CrossRef]
28. Hirsch, D.; Steinfeld, A. Radiative transfer in a solar chemical reactor for the co-production of hydrogen and carbon by thermal decomposition of methane. *Chem. Eng. Sci.* **2004**, *59*, 5771–5778. [CrossRef]
29. Klein, H.H.; Karni, J.; Ben-Zvi, R.; Bertocchi, R. Heat transfer in a directly irradiated solar receiver/reactor for solid-gas reactions. *Sol. Energy* **2007**, *81*, 1227–1239. [CrossRef]
30. Villafan-Vidales, H.I.; Arancibia-Bulnes, C.A.; Dehesa-Carrasco, U.; Romero-Paredes, H. Monte Carlo radiative transfer simulation of a cavity solar reactor for the reduction of cerium oxide. *Int. J. Hydrogen Energy* **2009**, *34*, 115–124. [CrossRef]
31. Z'Graggen, A.; Steinfeld, A. Hydrogen production by steam gasification of carbonaceous materials using concentrated solar energy—V. Reactor modeling, optimization, and scaleup. *Int. J. Hydrogen Energy* **2008**, *33*, 5484–5492. [CrossRef]
32. Martinek, J.; Bingham, C.; Weimer, A.W. Computational modeling and on-sun model validation for a multiple tube solar reactor with specularly reflective cavity walls. Part 1: Heat transfer model. *Chem. Eng. Sci.* **2012**, *81*, 298–310. [CrossRef]
33. Bellan, S.; Alonso, E.; Gomez-Garcia, F.; Perez-Rabago, C.; Gonzalez-Aguilar, J.; Romero, M. Thermal performance of lab-scale solar reactor designed for kinetics analysis at high radiation fluxes. *Chem. Eng. Sci.* **2013**, *101*, 81–89. [CrossRef]
34. Bellan, S.; Alonso, E.; Perez-Rabago, C.; Gonzalez-Aguilar, J.; Romero, M. Numerical modeling of solar thermochemical reactor for kinetic analysis. *Energy Procedia* **2014**, *49*, 735–742. [CrossRef]
35. Tapia, E.; González-Pardo, A.; Iranzo, A.; Vidal, A.; Rosa, F. Experimental testing of multi-tubular reactor for hydrogen production and comparison with a thermal CFD model. *AIP Conf. Proc.* **2018**, *2033*, 130013. [CrossRef]
36. Ballestrín, J.; Monterreal, R. Hybrid heat flux measurement system for solar central receiver evaluation. *Energy* **2004**, *29*, 915–924. [CrossRef]
37. González-Pardo, A.; Denk, T.; Vidal, A. Thermal Tests of a Multi-Tubular Reactor for Hydrogen Production by Using Mixed Ferrites Thermochemical Cycle. In *SolarPACES Conference Proceedings*; AIP Publishing: Abu Dhabi, UAE, 2016.
38. ANSYS Inc. *ANSYS CFX-Solver Theory Guide*; ANSYS: Canonsburg, PA, USA, 2011; p. 402.
39. Tapia, E.; Iranzo, A.; Pino, F.J.; Salva, A.; Rosa, F. Methodology for thermal design of solar tubular reactors using CFD techniques. *Int. J. Hydrogen Energy* **2016**, *41*, 19525–19538. [CrossRef]
40. Ma, T.; Ren, T.; Chen, H.; Zhu, Y.; Li, S.; Ji, G. Thermal performance of a solar high temperature thermochemical reactor powered by a solar simulator. *Appl. Therm. Eng.* **2019**, *146*, 881–888. [CrossRef]

© 2019 by the authors. Licensee MDPI, Basel, Switzerland. This article is an open access article distributed under the terms and conditions of the Creative Commons Attribution (CC BY) license (http://creativecommons.org/licenses/by/4.0/).

Article

Combining Microwave Pretreatment with Iron Oxide Nanoparticles Enhanced Biogas and Hydrogen Yield from Green Algae

Asad A. Zaidi [1], Ruizhe Feng [1], Adil Malik [1], Sohaib Z. Khan [2,4], Yue Shi [1,*], Asad J. Bhutta [1] and Ahmer H. Shah [3]

[1] College of Power and Energy Engineering, Harbin Engineering University, Harbin 150001, China; asadali@pnec.nust.edu.pk (A.A.Z.); charlesfrz@126.com (R.F.); adilmalik@hrbeu.edu.cn (A.M.); asadjaved@hrbeu.edu.cn (A.J.B.)
[2] Department of Mechanical Engineering, Faculty of Engineering, Islamic University of Madinah, Medina 42351, Saudi Arabia; szkhan@iu.edu.sa
[3] Department of Textile Engineering, Balochistan University of Information Technology, Engineering and Management Sciences, Quetta 87300, Pakistan; ahmer.shah@buitms.edu.pk
[4] Department of Engineering Sciences, PN Engineering College, National University of Sciences and Technology, Karachi 75350, Pakistan
* Correspondence: shiyue@hrbeu.edu.cn; Tel.: +86-185-4581-2660; Fax: +86-451-8251-9305

Received: 15 December 2018; Accepted: 2 January 2019; Published: 7 January 2019

Abstract: The available energy can be effectively upgraded by adopting smart energy conversion measures. The biodegradability of biomass can be improved by employing pretreatment techniques; however, such methods result in reduced energy efficiency. In this study, microwave (MW) irradiation is used for green algae (*Enteromorpha*) pretreatment in combination with iron oxide nanoparticles (NPs) which act as a heterogeneous catalyst during anaerobic digestion process for biogas enhancement. Batch-wise anaerobic digestion was carried out. The results showed that MW pretreatment and its combination with Fe_3O_4 NPs produced highest yields of biogas and hydrogen as compared to the individual ones and control. The biogas amount and hydrogen % v/v achieved by MW pretreatment + Fe_3O_4 NPs group were 328 mL and 51.5%, respectively. The energy analysis indicated that synergistic application of MW pretreatment with Fe_3O_4 NPs produced added energy while consuming less input energy than MW pretreatment alone. The kinetic parameters of the reaction were scientifically evaluated by using modified Gompertz and Logistic function model for each experimental case. MW pretreatment + Fe_3O_4 NPs group improved biogas production potential and maximum biogas production rate.

Keywords: algae; anaerobic digestion; biogas; biohydrogen; energy assessment; kinetic models; microwave; nanoparticles; pretreatment

1. Introduction

Anaerobic digestion (AD) is a microbial-mediated process which is widely used for the conversion of complex organic waste to renewable energy in the form of biogas [1]. The synergistic catalysis of various microorganisms without oxygen determines the biological route of the AD process. The organic matter conversion to biogas follow four main conversion phases namely; hydrolysis, acidogenesis, acetogenesis, and methanogenesis [2]. During the hydrolysis stage complex polymeric organic matter including carbohydrates, proteins, and fats transform into simple organic monomers by the action of hydrolytic bacteria. The monomers such as sugar, amino acids, and fatty acids are then converted into volatile fatty acids (VFAs) under the action of fermentative bacteria during the second stage called as acidogenesis. During the third phase, acetogenic bacteria transforms VFAs into acetic acid and hydrogen (H_2) gas. Methanogenic bacteria transform acetic acid and H_2 into methane (CH_4) and carbon dioxide (CO_2) [3]. The quality of biogas in terms of composition varies depending on biomass, precursors, additives and the conversion process. In general, biogas contains 50–75% methane and 25–45% carbon dioxide, in addition to small amounts of other gases and typically has a calorific value of 21–24 MJ/m^3 [4].

One of the potential feedstocks for biogas generation by AD process is algal biomass [5]. Algae are unicellular or multicellular organisms. In comparison with other biomass, they possess many benefits such as they can grow in natural and artificial systems, they can grow in fresh and marine water [6,7]. In addition, they have high biomass yield and greater carbon dioxide capture. Algal biomass can offer numerous biofuels such as biohydrogen, methane, biodiesel, bioethanol, and biogas [8]. The strong resistant algae cell wall is composed of three main components: biopolymers, cellulose, and hemicellulose. These components play a protective role in cells. Cellulose molecules are arranged regularly in the form of bundles. It also contains a small portion of pectin, protein, ash, and extracts, including soluble non-structural substances, non-structural sugars, nitrogen compounds, chlorophyll, and waxes [9]. However, the inter- and intra-molecular hydrogen bonds have made the dissolution of cellulose a difficult process in common solvents. This hinders or limits the anaerobic digestion of algal biomass during the hydrolysis stage.

Numerous pretreatment methods for algae can be used including biological (enzymatic), chemical (acid or alkali), physical (ultrasound, microwave, or shear force) and thermal methods [10]. However, selection of a pretreatment process is mainly reliant on its low capital cost, positive energy balance, and lesser operational cost to make AD process economically feasible [11]. Microwave (MW) pretreatment is the transmission of electromagnetic energy in the frequency range of 0.3 to 300 GHz. MW pretreatment involves no contact amongst the source and the chemicals [12]. Passos et al. [13] studied the effect of MW pretreatment on algae from High Rate Algal Ponds (HRAP). Results showed that MW pretreatment enhanced biogas production rate (25–75%) and successfully improved the digestibility of algal biomass. Several studies discussed MW pretreatment applied to waste activated sludge [14–19]. Almost all the studies reported an enhancement in sludge solubilization and biogas generation. In our previous study [20], optimization of MW pretreatment for an AD of *Enteromorpha* was carried out using response surface methodology. Results showed that 24.4 mL biogas/g dry algae was produced at the optimized MW pretreatment conditions after AD.

The concerns about expansion in the bioenergy sector during the past decade have driven a number of scientists and researchers to pursue innovative solutions for its production. Nanotechnology is one of the emerging branches of science. It deals with dimensions less than 100 nm. It is the art of manipulating individual atoms. It is the most striking and fertile field which allows researchers to work at the molecular level [21]. In the field of bioenergy, nanotechnology can be applied for feedstock modification and more efficient catalysis. Minerals are needed for microorganism development [4]. Liu et al. [22] reported that minerals deliver upright atmosphere for anaerobic bacteria inside a digester and enhance biogas and methane generation. In another study, Qiang et al. [23] stated that in the presence of iron, cobalt, and nickel, methanogenic bacteria grow quickly during enzyme production. Heavy metal ions such as Co, Cu, Fe, Mo, Ni, and Zn have been documented as essential for several reactions during AD by Luna-deRisco et al. [24]. Micronutrients such as Co, Ni, Fe, Mg, and Ca are crucial for a variety of chemical, biochemical, and microbiological reactions related to VFA utilization, biogas generation, and cell lysis [25]. Nanoparticles (NPs) of micronutrients had an augmented effect on biogas production. Cascals et al. [26] applied 100 ppm (100 mg/L) of Fe_3O_4 NPs (7 nm) to organic waste in an anaerobic digester under mesophilic conditions (37 °C) for 60 days. Results showed an enhancement of 180% in biogas and 234% increase in methane yield. The authors mentioned that Fe^{2+} act as a unique source, which disintegrates the organic matter and increases biogas production in the anaerobic bacterial reactor. Suanon et al. [27] studied the metal distribution conversion during AD of wastewater sludge under the presence of Fe_3O_4 NPs. Batch anaerobic system was used under mesophilic conditions (37 °C). Methane production increases by 1.25 and 0.9 times by

0.75 g and 1.5 g per 500 mL dose of Fe_3O_4 NPs, respectively. The addition of Fe_3O_4 NPs showed an improvement of metals stabilization in the digestate resulted in an enhancement of biogas and methane production. Abdelsalam et al. [28] examined the influence of Fe_3O_4 NPs with different concentrations (5, 10, and 20 mg/L) on biogas and methane yield from the AD of cattle manure (CM) slurry. Anaerobic fermentation of CM was carried out batch-wise at operating temperature and mixing rate of 37 ± 0.3 °C and 90 rpm for a hydraulic retention time (HRT) of 50 days. The study indicated that the addition of 20 mg/L Fe_3O_4 NPs increases biogas production by 1.66 times and methane production by 1.96 times. Our previous work [29] investigated the effect of Fe_3O_4 NPs on biogas yield from anaerobic digestion of green algae (*Enteromorpha*). Results showed that the 10 mg/L of Fe_3O_4 NPs cumulative increase in biogas production was 28%. It was observed that during the less effective domain NPs had no additional effect as a controlled sample. However, approximately after 60 h of the digestion process, NPs showed the incremental effect on biogas production. It has been suggested that combining the pretreatment with NPs may result in an early dissolution of the algae cell wall and provide faster action by NPs on stimulation of microorganisms to achieve high cumulative biogas yield with positive energy balance. Therefore, the objective of the present study is to examine the effect of combining a microwave (MW) pretreatment of *Enteromorpha* with Fe_3O_4 NPs. Energy ratio was calculated, and established prediction models are used to substantiate the experimental results of this work for the approximation of biogas generation during AD.

2. Materials and Methods

2.1. Raw Material

Anaerobic sludge was acquired from Harbin Wenchang Sewage Treatment Plant, Harbin, Heilongjiang province, China. Total suspension solids (TSS) of sludge were 6390 mg/L whereas Volatile Suspension Solids (VSS) were 2545 mg/L. The *Enteromorpha* was attained from the Institute of Hydrobiology of The Chinese Academy of Science, Wuhan, China. It was air-dried in the drying oven and then sealed in a bottle with a breathable film on the top. Each biodigester contained 60 mL of sludge and 20 g of Enteromorpha powder. The protein, fat and ash content of *Enteromorpha* were 13.20%, 1.06%, and 21.77%, respectively. Fe_3O_4 NPs (spherical shape with an average size < 100 nm were purchased from China Metallurgical Research Institute, Beijing, China. The concentration of NPs in the biomass was 10 mg/L. Similar NP concentration has been used in our previous study [29] and other studies [30,31]. In order to reduce the agglomeration of NPs, suspensions for the given concentration by adding distilled water containing sodium dodecylbenzene sulfonate (SDS) 0.1 mM was prepared [32].

2.2. Experimental Setup

The MW pretreatment was performed before AD. A household Panasonic microwave oven (1180 W) was used. The *Enteromorpha* solution was stirred after every minute. The MW pretreatment condition was liquid:solid, pretreatment time and pretreatment power of 20:1, 6 min and 600 W, respectively [20]. The batch-wise AD experiments were conducted through the anaerobic batch system. The laboratory glass bottles (working volume = 500 mL) were used as biodigesters and operated for 108 h. The biodigesters were airtight with rubber plugs. Nitrogen gas was purged through a digester for 5 min at the start to create anaerobic condition [33]. The environment inside digester has been retained at 37 °C [31] and 150 rpm mixing speed. The gas chromatography (SP-2100A, BFRL Co., Beijing, China) was employed to determine hydrogen content % (v/v) of the biogas. Thermogravimetric analysis (TGA) was conducted to explore the decomposition of algae intercellular organic compounds using TA Instruments Q50. TGA was performed at a heating rate of 20 °C/min from 40 to 600 °C under a constant nitrogen flow rate of 50 mL/min. A medical syringe with a long needle was used to collect the samples from air-tight bottles and transferred to small tubes covered with rubber stoppers to avoid gas loss. The biogas generation was measured twice a day whereas its composition was observed once. Each experiment was conducted in triplicate to reduce likely errors, and the average values are indicated. OriginPro 8 software was used to perform one-way ANOVA analysis of results, $p < 0.05$ was considered to be statistically significant.

2.3. Energy Balance Analysis

The energy assessment was evaluated via calculation of energy input needed for pretreatment and the enhancement in biohydrogen yield for pretreated *Enteromorpha* [11]. The input energy and output energy were calculated using Equations (1) and (2). The energy ratio (Equation (3)) was calculated as the energy output over energy input. If energy ratio value is greater than 1, it means that the energy yield from hydrogen generation during AD was higher in comparison with the energy required for MW pretreatment. It should be noted that this energy analysis does not include the energy required to dry biomass and other processes for precursors.

$$E_i = \frac{P \times t}{V \times TS} \quad (1)$$

where:

E_i = Energy input (kJ/gVS)
P = Power required for pretreatment (W)
t = Microwave pretreatment time (s)
V = Volume of biomass (L)
TS = Total solid in biomass (g TS/L)

$$E_o = \frac{\Delta P \times \varepsilon}{10^6} \quad (2)$$

where:

E_o = Energy output (kJ/gVS)
ΔP = Hydrogen yield (ml H_2/gVS)
ε = Calorific value of hydrogen (120,000 kJ/m^3)

$$\Delta E = \frac{E_o}{E_i} \quad (3)$$

2.4. Mathematical Kinetic Models

The AD process performance with the combined effect of MW pretreatment and Fe_3O_4 NPs was mathematically evaluated via modified Gompertz model Equation (4) [34] and Logistic Function model Equation (5) [35]. OriginPro 8 software was used to determine kinetic parameters for both models. The software uses an iterative method by employing the Levenberg-Marquardt (L-M) algorithm to estimate parameters for describing reaction kinetics. Akaike Information Criterion (AIC) test was performed to asses which model is better describing the kinetics of the AD process [36]. The model with lower AIC value suggests a better fit and predicting capability. For each model, the AIC value and Akaike's weight value was calculated by using Equations (6) and (7) [37]:

$$B = B_p \cdot \exp\left(-\exp\left(MBPR \frac{2.7183}{B_p} \cdot (BPDT - t) + 1\right)\right) \quad (4)$$

$$B = \frac{B_p}{1 + \exp\left[4MBPR \frac{BPDT - t}{B_p} + 2\right]} \quad (5)$$

where:

B = Cumulative biogas volume at digestion time t (mL)
B_P = Biogas production potential (mL)

MBPR = Maximum biogas production rate (mL/h)
BPDT = Biogas production delay time (h)
t = Total digestion time (h)

$$AIC = \begin{cases} N \ln \frac{RSS+2K}{N}, \text{ when } \frac{N}{K} \geq 40 \\ N \ln \frac{RSS}{N} + 2K + \frac{2K(K+1)}{N-K-1}, \text{ when } \frac{N}{K} < 40 \end{cases} \quad (6)$$

$$\text{Akaike's weight} = \frac{e^{-0.5\Delta AIC}}{1 + e^{-0.5\Delta AIC}} \quad (7)$$

where:
N = Number of points
RSS = Residual sum of square
K = Number of model parameters
ΔAIC = The relative difference between the two AIC values

3. Results and Discussion

3.1. Biogas and Hydrogen Production

Biogas production influenced by MW pretreatment and its combination with Fe_3O_4 NPs is shown in Figure 1. It is to be noted that all treatments improved the biogas production as compared to control. The maximum total biogas yield of 328 mL was achieved by MW pretreatment + Fe_3O_4 NPs group. The MW pretreatment and Fe_3O_4 NPs individually produced 302 and 289 mL, respectively. The Enteromorpha cell wall comprises an external layer and an internal layer. The external layer is an electron dense polymeric matrix in which glycoprotein and carbohydrates are present, whereas cellulose and hemicellulose exist in the internal layer [38]. During the initial stage, the increase in biogas in combined Fe_3O_4 NPs and MW pretreatment groups is credited to the pretreatment method. MW pretreatment rises the lysis rate which results in the increasing effect on biogas production [39]. MW pretreatment hydrolyzes the glycosidic bond present in carbohydrates and polysaccharides which turns into simple sugars. The dissolution of the algae cell wall by MW pretreatment can clearly be elucidated by the results shown in Figure 2 and Table 1. The TGA and Difference Thermo Gravimetry (DTG) graphs show better degradation of MW pretreated samples as compared to the control sample. The first mass loss region ranging from 50 °C to 200 °C corresponds to evaporation of moisture and degradation of organic species. As can be noted, MW pretreatment shows a smooth single peak at a temperature of 80 °C while the control sample shows small peaks at the temperature of 69 °C, 81 °C, and 96 °C. The first mass loss values of $T_{5\%}$ was decreased from 94 °C to 88 °C and the second stage $T_{10\%}$, increased from 183 °C to 196 °C for control and MW pretreated, respectively. Moreover, DTG is shown in Figure 2b, two peaks are showing the presence of hemicellulose and cellulose in the control sample at a temperature of 251 and 341 °C. It can be observed that MW pretreatment destroyed the hemicellulose to a greater extent making it available for anaerobic bacteria to produce biogas [40]. However, the peak height (max. rate of degradation) of cellulose peak for control at 341 °C is slightly affected and is shifted to 0.26 from 0.20%/°C due to MW pretreatment. This showed that MW destroyed the organic species and hemicellulose to a greater extent while the structure of cellulose was slightly altered and opens, which may account for increased biogas production. Similar results are reported for cellulose effects in the literature [41].

In a later stage, further dissolution of internal layer occurred by the attack of NPs. The hydrolysis of cellulose by NPs produce oligosaccharides such as cellobiose and cellodextrin [42]. The biopolymers (proteins, carbohydrates, and lipids) released by dissolution of the cell wall are then changed into amino acids, simple sugars, peptides and volatile fatty acids [40]. The maximum cumulative biogas and amount of hydrogen produced during the experiment are shown in Figure 3a,b. Fe_3O_4 NPs + MW pretreatment group produced the highest amount of biogas and highest hydrogen content (% v/v).

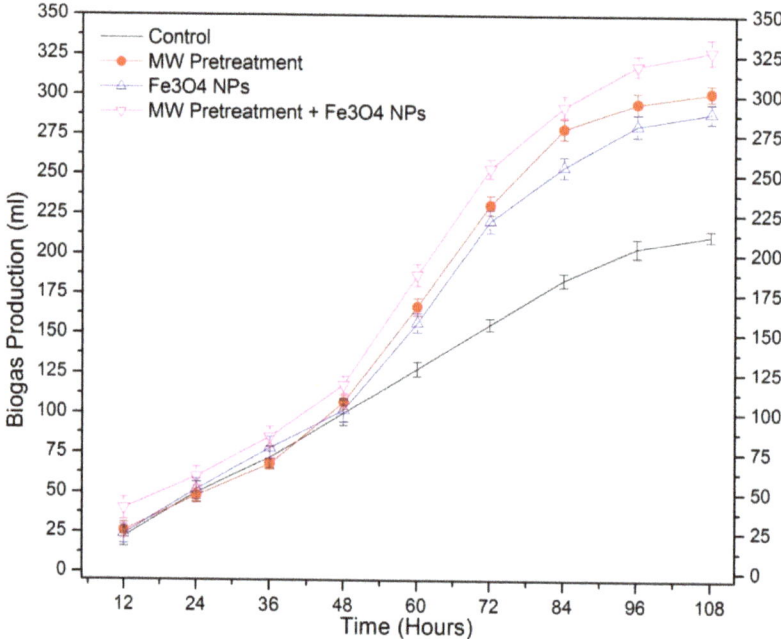

Figure 1. Biogas production influenced by microwave (MW) pretreatment and its combination with Fe_3O_4 nanoparticles (NPs).

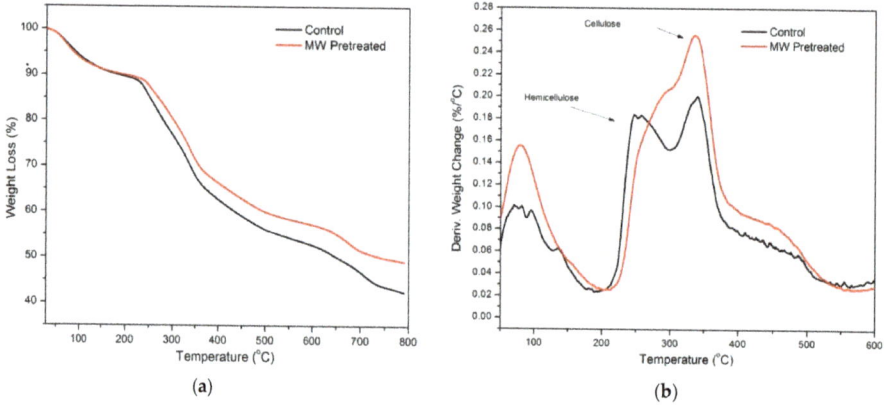

Figure 2. Thermogravimetric analysis (TGA) (**a**) and Difference Thermo Gravimetry (DTG) (**b**) of algae before and after MW pretreatment.

Table 1. Thermogravimetric Analysis (TGA) and Difference Thermo Gravimetry (DTG) results of *Enteromorpha* before and after microwave (MW) pretreatment.

Sample	$T_{5\%}$ (°C)	$T_{10\%}$ (°C)	Y_c (%) at 600 °C	Cellulose DTG Peak (°C)	Hemicellulose DTG Peak (°C)
Control	94	183	42	341	251
MW Pretreated	88	196	49	336	297

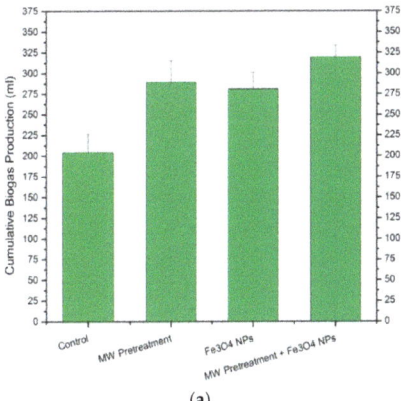
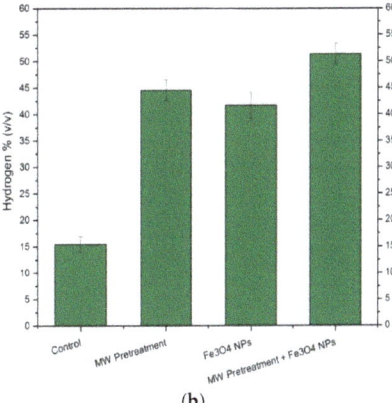

(a) (b)

Figure 3. Cumulative biogas production (**a**) and hydrogen % v/v by different treatment conditions.

Similar results have been obtained by Abdelsalam et al. [31]. The authors studied the influence of Fe_3O_4 NPs with AD of CM slurry. Biogas enhancement of 1.7 times than the control was reported. In another study, Suanon et al. [43] stated an enhancement of 1.27 times in biogas by Fe_3O_4 NPs. Cascals et al. [26] mentioned that Fe^{2+} act as a unique source, which disintegrates the organic matter and increases biogas production in the anaerobic bacterial reactor. According to Zhang and Lu [44], Fe_3O_4 NPs accelerate the reaction kinetics, increase biogas yield and reduce lag time. Our results are in agreement with Passos et al. [13] who stated an increased biogas production rate and a high degree of biomass solubilization by MW pretreatment of algae from HRAP. Zheng et al. [15] studied the effect of MW irradiance on primary sludge solubilization. The results showed that MW pretreatment improved soluble chemical oxygen demand (SCOD) in sludge and the biogas production was enhanced by 37%.

3.2. Energy Assessment

The energy generated (biohydrogen) from *Enteromorpha* AD for all groups (i.e., output energy, E_o) was calculated as shown in Table 2. The highest E_o (20.28 kJ/gVS) was achieved by MW pretreatment + Fe_3O_4 NPs group. For Fe_3O_4 NPs, MW pretreatment and Control groups, E_o amount of 14.45, 16.15, and 3.93 kJ/gVS was produced, respectively. This shows that all the treatments resulted in an increased output energy as compared to the control sample. Energy assessment of algal biomass AD process was conducted for estimating the feasibility of the MW pretreatment and its combined effect with Fe_3O_4 NPs. For this purpose, the output energy was divided by the energy needed for MW pretreatment (i.e., energy input, E_i) for MW pretreatment and MW pretreatment + Fe_3O_4 NPs groups. For both, the MW pretreatment + Fe_3O_4 NPs group and MW pretreatment alone, the energy ratio was higher than one. However, the energy ratio of combined effect is higher (i.e., 1.87) as compared to MW pretreatment alone (i.e., 1.49). This indicates that the enhancement in hydrogen production obtained was enough for covering the MW energy input to the AD system, which may be described by the spontaneity in the AD process after the applied treatments.

Table 2. Results for Energy Analysis.

	E_{in} (kJ/gVS)	E_{out} (kJ/gVS)	Energy Ratio
Control	-	3.93	-
MW Pretreatment	10.80	16.15	1.49
Fe_3O_4 NPs	-	14.45	-
MW Pretreatment + Fe_3O_4 NPs	10.80	20.28	1.87

3.3. Mathematical Kinetic Models

Kinetic parameters for the cumulative biogas produced by Enteromorpha AD were found out using modified Gompertz and Logistic Function models [34,35]. The results obtained from the kinetic study using the modified Gompertz and Logistic model are given in Tables 3 and 4, respectively. Figures 4 and 5 showed the contrast of predicted and experimental cumulative biogas yield by all groups. When applying the modified Gompertz model, maximum biogas production rate (MBPR) for control was 2.46 mL/h. For MW pretreatment, Fe_3O_4 NPs and MW pretreatment + Fe_3O_4 NPs, the MBPR found to be 4.32, 3.77, and 4.23 mL/h, respectively. Correspondingly, for the Logistic model, the maximum biogas production rate (MBPR) for the untreated, MW pretreatment, Fe_3O_4 NPs, and MW pretreatment + Fe_3O_4 NPs were 2.62, 4.87, 4.23, and 4.77 mL/h respectively. It is determined by both the kinetic models that combined effect of MW pretreatment and NPs had improved the biogas generation rate and reduced the lag phase time with respect to other groups. The decrease in lag phase was observed due to early hydrolysis of algae cell walls at the first stage of AD by MW pretreatment. This resulted in a faster consumption of sugar by anaerobic bacteria in later stages of AD. The correlation coefficient for the modified Gompertz model and Logistic Function model was above 98.01% and 99.18%, respectively. This suggests that both the models were fitting well with the experimental data. Table 5 shows the obtained results for the Akaike Information Criterion (AIC) test. AIC suggests that the modified Gompertz model has a lower AIC value and hence proved to be a better model to use in this case.

Table 3. Kinetic Parameters from the Modified Gompertz Model.

Parameter	Treatments			
	Control	MW Pretreatment	Fe_3O_4 NPs	MW Pretreatment + Fe_3O_4 NPs
B_p (mL)	268.11	374.09	374.528	426.354
MBPR (mL/h)	2.468	4.326	3.773	4.236
BPDT (h)	0.287	0.816	0.672	0.618
R^2	0.99728	0.98227	0.98457	0.98017
Predicted Biogas Yield (mL)	215.891	315.977	300.682	342.302
Measured Biogas Yield (mL)	212	302	289	328
Difference between measured and predicted biogas yield (%)	1.83	4.62	4.04	4.36

Remarks: B_p, Biogas production potential; MBPR, Maximum biogas production rate; BPDT, Biogas production delay time; R^2, Correlation Coefficient.

Table 4. Kinetic parameters from the Logistic Function Model.

Parameter	Treatments			
	Control	MW Pretreatment	Fe_3O_4 NPs	MW Pretreatment + Fe_3O_4 NPs
B_p (mL)	232.56	324.72	316.10	358.53
MBPR (mL/h)	2.628	4.870	4.230	4.771
BPDT (h)	0.443	1.023	0.887	0.839
R^2	0.99651	0.99414	0.99298	0.99184
Predicted Biogas Yield (mL)	213.244	309.394	295.084	335.453
Measured Biogas Yield (mL)	212	302	289	328
Difference between measured and predicted biogas yield (%)	0.58	2.44	2.10	2.27

Remarks: B_p, Biogas production potential; MBPR, Maximum biogas production rate; BPDT, Biogas production delay time; R^2, Correlation Coefficient.

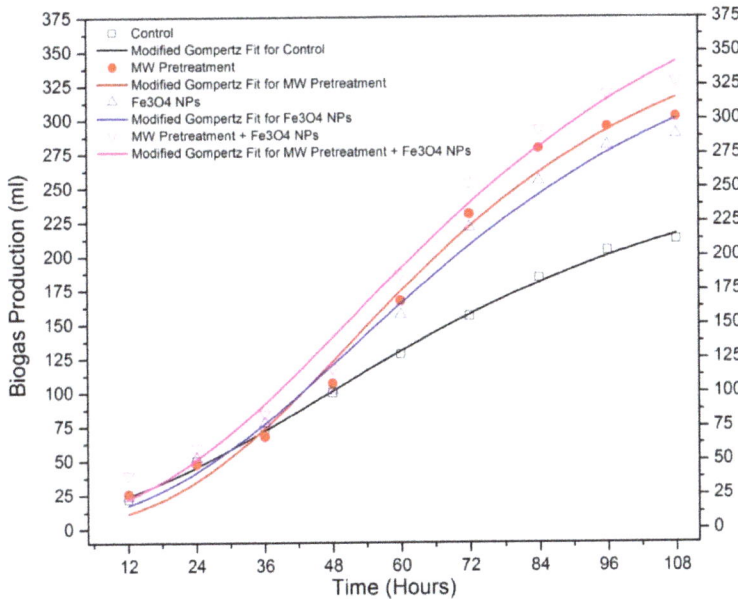

Figure 4. Modified Gompertz model fitting for experimental data.

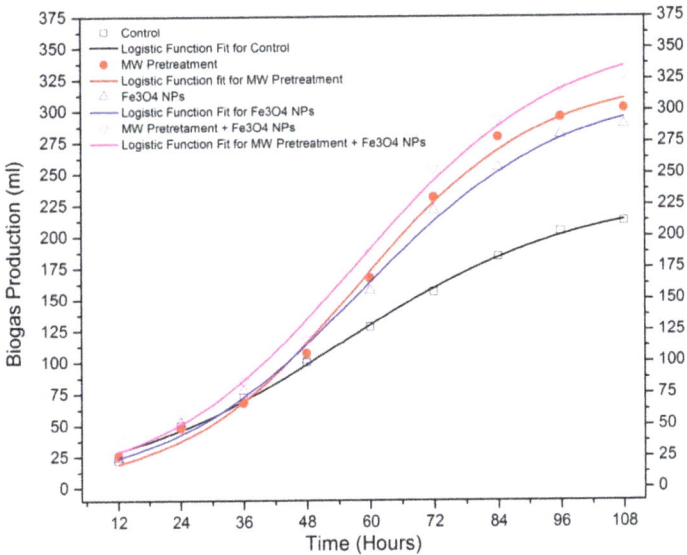

Figure 5. Logistic Function model fitting for experimental data.

Table 5. Akaike's information criterion (AIC) test results.

Model	RSS	N	AIC	Akaike Weight
Modified Gompertz Model	77.44757	9	37.37139	0.75284
Logistic Function Model	99.19747	9	39.59899	0.24716

Remarks: RSS, the Residual sum of the square; N, Number of Points; AIC, Akaike's Information Criterion.

4. Conclusions

The combined effect of MW pretreatment and Fe_3O_4 NPs showed improvement in biodegradability of green algae. The cumulative enhancement in biogas yield for MW pretreatment, Fe_3O_4 NPs and Fe_3O_4 NPs + MW pretreatment was 42.45%, 36.32%, and 54.71%, respectively. The energy assessment showed the high energy ratio of 1.87 is achieved by Fe_3O_4 NPs + MW pretreatment group. The experimental data of these results are further modeled via modified Gompertz and Logistic function model. Akaike Information Criterion (AIC) test highlighted that the modified Gompertz model is nearly matching with the experimental data. This study suggested that positive energy balance occurs when MW pretreatment is combined with Fe_3O_4 NPs for an AD of algal biomass. This study is applicable to all lignocellulose and other biomass with resistant cell walls or cellulose structure to improve the hydrolysis stage to produce a high amount of energy. The energy analysis indicates that combining MW pretreatment with small concentrations of Fe_3O_4 NPs causes added output energy. The results suggest an energy efficient way of producing biohydrogen and can easily be scaled-up for commercial-scale biohydrogen production. This aspect can produce fruit bearing results in the future production of biohydrogen via AD technology. In addition, cost-benefit analysis, optimization of process parameters, bioreactors design and more efficient energy conversion methods for biohydrogen could be the future scope of research for improved commercial and economic feasibility.

Author Contributions: All the authors contributed equally to the work. All authors have read and approved the final manuscript.

Funding: This research was financially supported by the National Key R&D Plan of China (2017YFC1404605), the Natural Science Foundation of China (Grant No. 51579049 and 51509044), the Fundamental Research Funds for the Central Universities (HEUCFG201820), the High Tech Ship Program (17KJ-003) and the Research on 130,000 GT cruise ship development.

Conflicts of Interest: The authors declare no conflict of interest.

Nomenclature

E_i	Energy input (kJ/gVS)
P	Power required for pretreatment (W)
t	Microwave pretreatment time (s)
V	Volume of biomass (L)
TS	Total solid in biomass (g TS/L)
E_o	Energy output (kJ/gVS)
ΔP	Hydrogen yield (mL H_2/gVS)
ε	Calorific value of hydrogen (120,000 kJ/m^3)
B	Cumulative biogas volume at digestion time t (mL)
B_P	Biogas production potential (mL)
MBPR	Maximum biogas production rate (mL/h)
BPDT	Biogas production delay time (h)
t	Total digestion time (h)
N	Number of points
RSS	Residual sum of square
K	Number of model parameters
ΔAIC	The relative difference between the two AIC values
MW	Microwave
AD	Anaerobic digestion
CO_2	Carbon dioxide
H_2	Hydrogen
H_2S	Hydrogen Sulfide
CH_4	Methane
VFAs	Volatile Fatty Acids
HRAP	High Rate Algal Ponds

CM	Cattle Manure
HRT	Hydraulic Retention Time
NPs	Nanoparticles
TSS	Total Suspension Solids
VSS	Volatile Suspension Solids
SDS	Sodium Dodecylbenzene Sulfonate
TGA	Thermogravimetric analysis
(L-M)	Levenberg-Marquardt
AIC	Akaike Information Criterion
SCOD	Soluble Chemical Oxygen Demand
DTG	Difference Thermo Gravimetry

References

1. Zhang, Q.; Hu, J.; Lee, D.J. Biogas from anaerobic digestion processes: Research updates. *Renew. Energy* **2016**, *98*, 108–119. [CrossRef]
2. Panico, A.; Antonio, G.; Esposito, G.; Frunzo, L.; Iodice, P.; Pirozzi, F. The Effect of Substrate-Bulk Interaction on Hydrolysis Modeling in Anaerobic Digestion Process. *Sustainability* **2014**, *6*, 8348–8363. [CrossRef]
3. Mao, C.; Feng, Y.; Wang, X.; Ren, G. Review on research achievements of biogas from anaerobic digestion. *Renew. Sustain. Energy Rev.* **2015**, *45*, 540–555. [CrossRef]
4. Ganzoury, M.A.; Allam, N.K. Impact of nanotechnology on biogas production: A mini-review. *Renew. Sustain. Energy Rev.* **2015**, *50*, 1392–1404. [CrossRef]
5. Solé-Bundó, M.; Salvadó, H.; Passos, F.; Garfí, M.; Ferrer, I. Strategies to Optimize Microalgae Conversion to Biogas: Co-Digestion, Pretreatment and Hydraulic Retention Time. *Molecules* **2018**, *23*, 2096. [CrossRef] [PubMed]
6. Bharathiraja, B.; Chakravarthy, M.; Ranjith Kumar, R.; Yogendran, D.; Yuvaraj, D.; Jayamuthunagai, J.; Praveen Kumar, R.; Palani, S. Aquatic biomass (algae) as a future feed stock for bio-refineries: A review on cultivation, processing and products. *Renew. Sustain. Energy Rev.* **2015**, *47*, 634–653. [CrossRef]
7. González-Fernández, C.; Sialve, B.; Bernet, N.; Steyer, J.P. Comparison of ultrasound and thermal pretreatment of Scenedesmus biomass on methane production. *Bioresour. Technol.* **2012**, *110*, 610–616. [CrossRef] [PubMed]
8. Ghimire, A.; Kumar, G.; Sivagurunathan, P.; Shobana, S.; Saratale, G.D.; Kim, H.W.; Luongo, V.; Esposito, G.; Munoz, R. Bio-hythane production from microalgae biomass: Key challenges and potential opportunities for algal bio-refineries. *Bioresour. Technol.* **2017**, *241*, 525–536. [CrossRef]
9. Satyanarayana, K.G.; Mariano, A.B.; Vargas, J.V.C. A review on microalgae, a versatile source for sustainable energy and materials. *Int. J. Energy Res.* **2011**, *35*, 291–311. [CrossRef]
10. Córdova, O.; Santis, J.; Ruiz-Fillipi, G.; Zuñiga, M.E.; Fermoso, F.G.; Chamy, R. Microalgae digestive pretreatment for increasing biogas production. *Renew. Sustain. Energy Rev.* **2018**, *82*, 2806–2813. [CrossRef]
11. Córdova, O.; Passos, F.; Chamy, R. Physical Pretreatment Methods for Improving Microalgae Anaerobic Biodegradability. *Appl. Biochem. Biotechnol.* **2018**, *185*, 114–126. [CrossRef]
12. Appels, L.; Houtmeyers, S.; Degrève, J.; Van Impe, J.; Dewil, R. Influence of microwave pre-treatment on sludge solubilization and pilot scale semi-continuous anaerobic digestion. *Bioresour. Technol.* **2013**, *128*, 598–603. [CrossRef]
13. Passos, F.; Solé, M.; García, J.; Ferrer, I. Biogas production from microalgae grown in wastewater: Effect of microwave pretreatment. *Appl. Energy* **2013**, *108*, 168–175. [CrossRef]
14. Climent, M.; Ferrer, I.; del Mar Baeza, M.; Artola, A.; Vázquez, F.; Font, X. Effects of thermal and mechanical pretreatments of secondary sludge on biogas production under thermophilic conditions. *Chem. Eng. J.* **2007**, *133*, 335–342. [CrossRef]
15. Zheng, J.; Kennedy, K.J.; Eskicioglu, C. Effect of low temperature microwave pretreatment on characteristics and mesophilic digestion of primary sludge. *Environ. Technol.* **2009**, *30*, 319–327. [CrossRef] [PubMed]
16. Sólyom, K.; Mato, R.B.; Pérez-Elvira, S.I.; Cocero, M.J. The influence of the energy absorbed from microwave pretreatment on biogas production from secondary wastewater sludge. *Bioresour. Technol.* **2011**, *102*, 10849–10854. [CrossRef]

17. Eskicioglu, C.; Kennedy, K.J.; Droste, R.L. Characterization of soluble organic matter of waste activated sludge before and after thermal pretreatment. *Water Res.* **2006**, *40*, 3725–3736. [CrossRef]
18. Park, W.-J.; Ahn, J.-H.; Hwang, S.; Lee, C.-K. Effect of output power, target temperature, and solid concentration on the solubilization of waste activated sludge using microwave irradiation. *Bioresour. Technol.* **2010**, *101*, S13–S16. [CrossRef]
19. Eskicioglu, C.; Kennedy, K.J.; Droste, R.L. Enhanced disinfection and methane production from sewage sludge by microwave irradiation. *Desalination* **2009**, *248*, 279–285. [CrossRef]
20. Feng, R.; Zaidi, A.A.; Zhang, K.; Shi, Y. Optimisation of Microwave Pretreatment for Biogas Enhancement through Anaerobic Digestion of Microalgal Biomass. *Period. Polytech. Chem. Eng.* **2018**, *63*, 65–72. [CrossRef]
21. Antonio, F.; Antunes, F.; Gaikwad, S.; Ingle, A.P. Nanotechnology for Bioenergy and Biofuel Production. In *Green Chemistry and Sustainable Technology*; Springer: Berlin, Germany, 2017; pp. 3–18. ISBN 978-3-319-45458-0.
22. Liu, L.; Zhang, T.; Wan, H.; Chen, Y.; Wang, X.; Yang, G.; Ren, G. Anaerobic co-digestion of animal manure and wheat straw for optimized biogas production by the addition of magnetite and zeolite. *Energy Convers. Manag.* **2015**, *97*, 132–139. [CrossRef]
23. Qiang, H.; Niu, Q.; Chi, Y.; Li, Y. Trace metals requirements for continuous thermophilic methane fermentation of high-solid food waste. *Chem. Eng. J.* **2013**, *222*, 330–336. [CrossRef]
24. Luna-delRisco, M.; Orupõld, K.; Dubourguier, H.C. Particle-size effect of CuO and ZnO on biogas and methane production during anaerobic digestion. *J. Hazard. Mater.* **2011**, *189*, 603–608. [CrossRef] [PubMed]
25. Menon, A.; Wang, J.; Giannis, A. Optimization of micronutrient supplement for enhancing biogas production from food waste in two-phase thermophilic anaerobic digestion. *Waste Manag.* **2017**, *59*, 465–475. [CrossRef] [PubMed]
26. Casals, E.; Barrena, R.; Garcia, A.; Gonzalez, E.; Delgado, L.; Busquets-Fite, M.; Font, X.; Arbiol, J.; Glatzel, P.; Kvashnina, K.; et al. Programmed iron oxide nanoparticles disintegration in anaerobic digesters boosts biogas production. *Small* **2014**, *10*, 2801–2808. [CrossRef]
27. Suanon, F.; Sun, Q.; Mama, D.; Li, J.; Dimon, B.; Yu, C.P. Effect of nanoscale zero-valent iron and magnetite (Fe3O4) on the fate of metals during anaerobic digestion of sludge. *Water Res.* **2016**, *88*, 897–903. [CrossRef]
28. Abdelsalam, E.; Samer, M.; Attia, Y.A.; Abdel-Hadi, M.A.; Hassan, H.E.; Badr, Y. Influence of zero valent iron nanoparticles and magnetic iron oxide nanoparticles on biogas and methane production from anaerobic digestion of manure. *Energy* **2017**, *120*, 842–853. [CrossRef]
29. Zaidi, A.A.; RuiZhe, F.; Shi, Y.; Khan, S.Z.; Mushtaq, K. Nanoparticles augmentation on biogas yield from microalgal biomass anaerobic digestion. *Int. J. Hydrogen Energy* **2018**, *43*, 14202–14213. [CrossRef]
30. Abdelsalam, E.; Samer, M.; Attia, Y.A.; Abdel-Hadi, M.A.; Hassan, H.E.; Badr, Y. Effects of Co and Ni nanoparticles on biogas and methane production from anaerobic digestion of slurry. *Energy Convers. Manag.* **2017**, *141*, 108–119. [CrossRef]
31. Abdelsalam, E.; Samer, M.; Attia, Y.A.; Abdel-hadi, M.A.; Hassan, H.E.; Badr, Y. Comparison of nanoparticles effects on biogas and methane production from anaerobic digestion of cattle dung slurry. *Renew. Energy* **2016**, *87*, 592–598. [CrossRef]
32. Khan, S.Z.; Yuan, Y.; Abdolvand, A.; Schmidt, M.; Crouse, P.; Li, L.; Liu, Z.; Sharp, M.; Watkins, K.G. Generation and characterization of NiO nanoparticles by continuous wave fiber laser ablation in liquid. *J. Nanopart. Res.* **2009**, *11*, 1421–1427. [CrossRef]
33. Xia, A.; Jacob, A.; Tabassum, M.R.; Herrmann, C.; Murphy, J.D. Production of hydrogen, ethanol and volatile fatty acids through co-fermentation of macro- and micro-algae. *Bioresour. Technol.* **2016**, *205*, 118–125. [CrossRef] [PubMed]
34. Syaichurrozi, I.; Budiyono; Sumardiono, S. Predicting kinetic model of biogas production and biodegradability organic materials: Biogas production from vinasse at variation of COD/N ratio. *Bioresour. Technol.* **2013**, *149*, 390–397. [CrossRef] [PubMed]
35. Deepanraj, B.; Sivasubramanian, V.; Jayaraj, S. Effect of substrate pretreatment on biogas production through anaerobic digestion of food waste. *Int. J. Hydrogen Energy* **2017**, *42*, 26522–26528. [CrossRef]
36. Rawlings, J.O.; Pantula, S.G.; Dickey, D.A. *Applied Regression Analysis: A Research Tool*, 2nd ed.; Springer: New York, NY, USA, 1998.
37. Wong, C.S.; Li, W.K. A note on the corrected Akaike information criterion for threshold autoregressive models. *J. Time Ser. Anal.* **1998**, *19*, 113–124. [CrossRef]

38. Demuez, M.; Mahdy, A.; Tomás-Pejó, E.; González-Fernández, C.; Ballesteros, M. Enzymatic cell disruption of microalgae biomass in biorefinery processes. *Biotechnol. Bioeng.* **2015**, *112*, 1955–1966. [CrossRef] [PubMed]
39. Mahdy, A.; Mendez, L.; Ballesteros, M.; González-Fernández, C. Enhanced methane production of Chlorella vulgaris and Chlamydomonas reinhardtii by hydrolytic enzymes addition. *Energy Convers. Manag.* **2014**, *85*, 551–557. [CrossRef]
40. Kavitha, S.; Yukesh Kannah, R.; Yeom, I.T.; Do, K.-U.; Banu, J.R. Combined thermo-chemo-sonic disintegration of waste activated sludge for biogas production. *Bioresour. Technol.* **2015**, *197*, 383–392. [CrossRef]
41. Shah, A.H.; Li, X.; Xu, X.D.; Wang, S.; Bai, J.W.; Wang, J.; Liu, W.B.; Sciences, M. Effect of alkali treated walnut shell (Juglansregia) on high performance thermosets, study of curing behavior, thermal and thermomechanical properties. *Dig. J. Nanomater. Biostruct.* **2018**, *13*, 857–873.
42. Ehimen, E.A.; Holm-Nielsen, J.-B.; Poulsen, M.; Boelsmand, J.E. Influence of different pre-treatment routes on the anaerobic digestion of a filamentous algae. *Renew. Energy* **2013**, *50*, 476–480. [CrossRef]
43. Suanon, F.; Sun, Q.; Li, M.; Cai, X.; Zhang, Y.; Yan, Y.; Yu, C.P. Application of nanoscale zero valent iron and iron powder during sludge anaerobic digestion: Impact on methane yield and pharmaceutical and personal care products degradation. *J. Hazard. Mater.* **2017**, *321*, 47–53. [CrossRef]
44. Zhang, J.; Lu, Y. Conductive Fe_3O_4 nanoparticles accelerate syntrophic methane production from butyrate oxidation in two different lake sediments. *Front. Microbiol.* **2016**, *7*, 1316. [CrossRef]

© 2019 by the authors. Licensee MDPI, Basel, Switzerland. This article is an open access article distributed under the terms and conditions of the Creative Commons Attribution (CC BY) license (http://creativecommons.org/licenses/by/4.0/).

MDPI
St. Alban-Anlage 66
4052 Basel
Switzerland
Tel. +41 61 683 77 34
Fax +41 61 302 89 18
www.mdpi.com

Processes Editorial Office
E-mail: processes@mdpi.com
www.mdpi.com/journal/processes

www.ingramcontent.com/pod-product-compliance
Lightning Source LLC
LaVergne TN
LVHW070640100526
838202LV00013B/846